The Use of [illegible] for Technical Students

R. A. KELLY B.A.

Second Edition
Metricated where appropriate

GEORGE G. HARRAP & CO. LTD
London Toronto Wellington Sydney

First published in Great Britain 1962
by GEORGE G. HARRAP & CO. LTD
182–184 High Holborn, London WC1V 7AX

Reprinted: 1963; 1965
Second Edition, metricated where appropriate, 1970

Reprinted 1971

ISBN 0 245 50843 0

Reproduced and Printed by
Redwood Press Limited
Trowbridge & London

PREFACE

This text book is intended to provide suitable exercises in technical and scientific exposition. It is based on the simple assumption that the young technologist needs to be able to use language functionally and to express himself clearly and lucidly within his own specialized subject and in subjects closely allied to it.

I make no apology for the absence of the 'literary' type of exercise. While I realize quite well that engineers are human, that they enjoy literature and are not blind to the beauties of our language, I also realize how great is their need for a guide to exposition—something to help them to escape from that hybrid literary/scientific writing which they often bring with them to a technical course.

In the longer exercises, however, I have tried to avoid too restricted an approach, and I would impress upon all students that they should not regard the type of exposition in this book as the horizon of their English studies, but as a practical and immediate objective.

Although the book was originally designed for technological students, the range of the exercises has been increased in an attempt to meet much of the demand of students at Ordinary National Certificate level and at some stages of the City and Guilds courses.

Above all, these exercises are offered as an aid to those English lecturers who are confronted—often suddenly—with groups of technical students and told to "take English with them." May this small contribution enable them to come more easily to terms with their new environment.

R. A. K.

Note on the Exercises

In certain sections of the book, the exercises are marked A and B. (A—Elementary exercises; B—More advanced exercises.) In other instances the exercises are graded in order of difficulty.

PREFACE TO THE SECOND EDITION

In preparing this edition SI metric units have been adopted where appropriate. Imperial units have been retained in the extracts and in questions relating to the extracts, since students are expected to be familiar with both systems during the transitional period. For the same reason certain non-SI metric terms —*e.g.*, the dyne—have also been retained.

R.A.K.

ACKNOWLEDGMENTS

For permission to use copyright extracts the author and publishers offer their grateful thanks to the following:

Messrs George Allen and Unwin, Ltd, for an extract from *Mathematics for the Million*, by Lancelot Hogben.

Messrs Edward Arnold, Ltd, for three extracts from Part One of *Workshop Technology*, by W. A. J. Chapman.

The *Autocar* and Mr D. M. Peters, for an extract from an article on *Brakes*.

Messrs G. Bell and Sons, Ltd, and Professor Andrade for an extract from *An Approach to Modern Physics*.

The British Petroleum Co., Ltd, for an extract from *The Search for Oil*.

The journal *Business*, for an extract from an article *Ergonomics*, by P. Bennet.

The Cambridge University Press, for an extract from Volume One of *The Major Achievements of Science*, by A. E. E. McKenzie, and another from *How Animals Move*, by Sir James Gray.

The Clarendon Press, Oxford, for an extract from Volume V of *A History of Technology*.

The Economist, for an extract from an article entitled *Safety in the Air*, published November 1, 1959.

Messrs Victor Gollancz, Ltd, for an extract from *Brighter than 1000 Suns*, by Dr Robert Jungk.

The Controller of H.M. Stationery Office, for permission to reproduce an extract from the 1954 edition of *The Admiralty Manual of Navigation*, entitled *The Dutchman's Log*, and an extract from *Choice of Careers Booklet No. 78: Engineering Work for Boys*.

Messrs Heywood and Co., Ltd, for an extract from *Basic Electronics*, by J. Daly and R. A. Greenfield.

David Higham Associates, Ltd, for an extract from *Famous British Engineers*, by L. Halward.

Messrs Hodder and Stoughton, Ltd, for an extract from *The Ascent of Everest*, by John Hunt, based on the original despatches from Brigadier Sir John Hunt and other members of the Everest Expedition to *The Times*, and published by Hodder and Stoughton.

Messrs Longmans, Green and Co. for extracts from *Heat Engines*, by A. C. Walshaw, *Engineering Material, Machine Tools and Processes*, by W. Steeds and *Marc Isambard Brunel*, a biography by L. T. C. Rolt.

Messrs Frederick Muller, Ltd, for an extract from *Jet: The Story of a Pioneer*, by Sir Frank Whittle.

The New Scientist and Sir Graham Sutton, C.B.E., D.Sc., F.R.S., for an extract from an article on *Tornadoes*.

The Oxford University Press for an extract from *Plastics To-day*, by A. Allcott.

Messrs Routledge and Kegan Paul, Ltd, for an extract from *The Social History of Lighting*, by William O'Dea.

Temple Press Books for an extract from *Interplanetary Flight*, by Arthur Clark.

The Times Publishing Co., Ltd, for an extract from *The Times* of October 6, 1930.

The author would also like to express his appreciation of the help and advice given by former colleagues at Kingston College of Technology, and to record his thanks to Mr Geoffrey Allen for assistance given in the final stages of the manuscript.

CONTENTS

I

SENTENCES

THE FUNCTION OF A SENTENCE

1. The Purpose of a Sentence

As we look about us we see objects in relation to their background or to other objects—a jet plane against sky and cloud, machines operating in a workshop, a bulldozer uprooting trees—and much of our communication is aimed at passing such experience to others by means of speech or writing.

Words are useful labels for the objects we see, but they mean little until they are placed within the framework of a sentence, and it is the sentence which is really the basic unit of expression.

Sentences should be complete in themselves. They may be in the form of:

(*a*) *a plain statement*—He pressed the switch.
(*b*) *a question*—Which part of the machine is defective?
(*c*) *a command, order, or instruction*—Set up that bar between centres.
(*d*) *a hope or wish*—I hope the replacement cable will arrive soon.
(*e*) *an emotional comment*—Bless my soul!

2. The Structure of a Sentence

The commonest rule in sentence structure is that every sentence must begin with a capital letter and end with a full stop. Unfortunately too many students consider that they need do nothing else, and many shapeless, incomplete, and meaningless collections are made into 'sentences' in this way. To write the collection of words "With reference to the delivery of the new machines" does not make a sentence. Something is missing. The reader is introduced to the topic, then left in the air just as he expects something to be said about the delivery. The sentence could be completed:

> With reference to the delivery of the new machines, I must remind you that the cost of transport will be extra.

Most sentences are designed to relate objects to other objects by action, and this function is particularly important in technical writing. The engineer constantly faces problems involving objects and relative movements.

> Ball bearings may be lubricated with grease.
> The operator feeds the tool along at a predetermined rate.
> The belt runs in 40° vee grooves.

Each of these sentences deals in some way with objects and movement, and expresses them in a simple manner, the words almost parallel to the actions they describe.

Sentences are constructed according to the complexity of the message being expressed and may vary from a simple statement to a complicated explanation:

(*a*) The feed-shaft of a simple lathe may be driven by a belt.
(*b*) If the feed-shaft of a simple lathe is driven by a belt there is a limit to the number of different feeds.
(*c*) The form of drive to the feed-shaft may vary but the lead-screw must be gear-driven.
(*d*) If the feed-shaft of a simple lathe is driven by a belt there is a limit to the number of different feeds, but the operation of the machine is safer because the belt will merely slip off if trouble occurs.

The first sentence makes a plain statement, the second adds a condition 'if . . .,' the third makes two parallel statements about the feed-shaft and the lead-screw, and the last sentence contains simple statement, condition, reason, and parallel statements.

The terms we give to these types of sentences are:

(*a*) *Simple sentence*—a plain statement with a few modifications.
(*b*) *Complex sentence*—more detailed modifications and qualifications. Greater emphasis on subsidiary ideas.

(*c*) *Compound sentence*—parallel statements showing simple relationship within a single sentence.
(*d*) *Multiple sentence*—parallel ideas with detailed modifications and qualifications.

More important than knowing what these sentences are called is knowing that the framework can be used to give clarity and variety to your sentence structure.

3. Word Order in a Sentence

Always remember that the most important person in the communicating of ideas is the reader. He cannot read the writer's mind and depends solely on the words he receives; if these are not clearly arranged there is every chance that the message will not be received.

In technical writing it is important for the information to be presented quickly and clearly, and it is customary for sentences to reveal their subjects quickly.

> Water boils at 100° C.
>
> *not*
>
> In the case of water the boiling point is 100° C.

Help your reader to grasp the theme quickly.

Exercises

1. Indicate which of the following collections of words are sentences. Give your reasons.

A.

(*a*) A small engine drives the lifting gear.
(*b*) Even if the machine is lubricated regularly.
(*c*) The clamp holds the bar firmly.
(*d*) Having drilled a hole in the casting.
(*e*) He covered the bare wire with insulation tape.
(*f*) Off you go.
(*g*) The water flowing out through the waste pipe.
(*h*) When you consider what has happened.
(*i*) How should this adjustment be made?
(*j*) How to do it.
(*k*) Having repaired the amplifier.

(*l*) Under the bench.
(*m*) When he arrived.
(*n*) Where is the screwdriver?
(*o*) Take this message, please.

B.

(*a*) Beyond the Earth's galaxy lie further, distant galaxies.
(*b*) Where they found it.
(*c*) The main road, running for seven miles through flat countryside.
(*d*) May your new sales venture be a great success.
(*e*) Where he was going.
(*f*) Although the oil is low in viscosity we have no other lubricant to use.
(*g*) If the terms are suitable and if the guarantee given by the manufacturer is sufficient to meet our demands.
(*h*) Having discovered what we are to enter on the forms.
(*i*) In spite of the danger involved, Smith is determined to test that faulty machine.
(*j*) Whom we want to appoint and who is likely to appeal to the interviewing board.
(*k*) Ouch!
(*l*) Supposing the material does not arrive.
(*m*) The increase of absenteeism and the startling rise in the number of unofficial strikes during the last three months.
(*n*) Contrary to expectations the situation was a quiet one.
(*o*) Releasing the man for overseas work was a brilliant idea.

2. Indicate the type of sentence—Simple, Complex, Compound, or Multiple.

(*a*) He placed the beaker on the scales.
(*b*) y is a function of $2xy$.
(*c*) Tell me when the job is finished.
(*d*) He took the white metal bearing which he had skimmed and attempted to fit it to the shaft.
(*e*) He scraped the metal carefully even though he knew the measurements he had made were hopelessly inaccurate.
(*f*) The bronzes, although fairly hard, can be machined quite easily, and the chips come off in short pieces.
(*g*) He set up the material in the 4-jaw chuck and adjusted the speed to turn at 160 rev/min.
(*h*) Hammer heads are made from a plain carbon steel and are shaped by stamping or forging.

(*i*) Spring callipers are easily adjusted, but equally good work is possible with firm join callipers.

(*j*) Although a geared connexion between two shafts gives a solid drive it is not as flexible as a belt drive, and is therefore more troublesome.

(*k*) Despite the interference from the steel framework of the building, the reception from the small portable was reasonable although not up to the standard of the larger mains radio.

(*l*) Having interviewed three candidates and being dissatisfied with the qualifications of each one, the board decided not to make the appointment.

(*m*) If the calculations are correct the beam will withstand a loading of 87·5 N/mm.

(*n*) Four-wheel drive often involves the use of three differentials but direct drive to each wheel can overcome this difficulty if the system is a simple one.

(*o*) Having arranged for the servicing to be done during the holiday, he ran the machines above their usual number of hours and found that troubles arose, the majority concerning lubrication and wear of moving parts.

ADVICE ON SENTENCE CONSTRUCTION

A sentence should be accurate, consistent, direct, methodical, and interesting.

BE ACCURATE (*a*) Write grammatically
(*b*) Verify your information

Do not: make errors of fact
make careless grammatical errors
misquote information
exaggerate
make hasty generalizations
be vague (especially in using *rather* or *quite*)

e.g. (*a*) It is important for you to treat these kind of machines with care.
Grammatical error—*these* is plural; *kind* is singular, and the sentence should read:
It is important for you to treat this kind of machine (or these kinds of machine) with care.

(*b*) The camera has a reasonable lens and quite a good range of camera speeds.
This is too vague. You should give fuller information.
The camera has an f2.8 50 mm. lens, and a 5-speed shutter ranging from 1 second to 1/500th of a second.

Exercises

1. The following sentences contain errors of grammar. Some are obvious; others may not be as easy to find. Look particularly for: errors of agreement; confusion between pronouns; wrong use of prepositions; unrelated participles and relative pronouns. Write a corrected version of each sentence.

A.

(*a*) It was the foreman who he gave the work to.
(*b*) The tools, which included a chisel, mallet, marking gauge, and scriber, was placed in the tool chest.
(*c*) Having switched on the current the lighting was available.
(*d*) The operation of this machine is different than that of the other.
(*e*) Neither of the bonus and incentive schemes were acceptable to the works committee.
(*f*) There were less members at the meeting than we had expected.
(*g*) Each of the machines are to be overhauled during the holiday.
(*h*) He cut the metal the same width with the last strip.
(*i*) He asked him whom he thought was responsible for the breakage.
(*j*) When the workmen are keen and fit they work like any other craftsmen do.
(*k*) The main supplies of the ore comes from Scandinavia.
(*l*) Every man in the workshop is expected to think for themselves.
(*m*) He tried to carefully steady the drill.
(*n*) The foreman said that he will see that the safety regulation is observed.
(*o*) This is one of the best designs which has been presented to the committee.
(*p*) He could not find in the test-tube hardly any trace of the crystals he expected.
(*q*) The two meter readings conflicted, but he used the one what he considered most accurate.

(*r*) Having passed a current through the wire it is heated and expands.
(*s*) The lathe with various attachments are able to deal with several jobs.
(*t*) Jack told John that he was entitled to three days' sickness benefit.
(*u*) Heat may be transmitted both by conduction and convection.
(*v*) Using a piece of litmus paper an acid can be detected.

B.

(*a*) A lathe is basically a machine for turning between centres, but they may be used for other jobs.
(*b*) One should always pay careful attention to the operation of safety devices on machines or you will probably have an accident unless you are not careful.
(*c*) By using closed-circuit television in the reactor solves the problem of investigating without risk of contamination.
(*d*) The foreman was doubtful about him being able to operate the machine properly.
(*e*) The float in the carburetter was punctured which means that the chamber flooded.
(*f*) This particular metal is as strong, if not stronger, than the metal used previous.
(*g*) He noticed many obsolete machines walking through the factory.
(*h*) The first of the air-conditioning processes is that it is drawn through a sticky filter to remove dust and dirt.
(*i*) If bus bars are air-insulated they must be set up so that it can stand the mechanical forces of short-circuiting—better have bitumen insulation.
(*j*) They constructed the road parallel with the railway so as to save space.
(*k*) The new machine is one of those kind of machines which is subject to the usual teething troubles.
(*l*) There were three methods of supplying the oil and the sales manager said he preferred the latter one.
(*m*) The new milling machine is more efficient than any equipment in the workshop.
(*n*) If anyone requires medical attention they might obtain it at the works clinic.
(*o*) The effect of waves, which is an important aspect of scientific study, are useful to both designers of ships, aircraft, and space ships.

2. The following sentences are examples of vague or inaccurate expression. Study them carefully; make sure you know what is wrong; then write a corrected version.

A.

(*a*) There are many things that can be done on a lathe.
(*b*) Pig-iron is 100% iron.
(*c*) When the machine was tested it was found that its efficiency was rather low.
(*d*) America was first to put a satellite into space.
(*e*) Shop-stewards are professional trouble makers.
(*f*) The circumference of the earth is quite big compared with the moon.
(*g*) The tailstock on a lathe is rather a difficult fixture to describe.
(*h*) The cross slide moves up and down the bed ways.
(*i*) When taking a big cut with a lathe tool you need a lot of coolant.
(*j*) I have not been able to start this car; there must be something wrong with it.
(*k*) A motor car needs oiling now and then.
(*l*) A micrometer is useful for working out the distances between points or across a piece of metal.
(*m*) The mechanical garden cultivator is rather a useful machine but if you do not maintain it properly it will not do so.
(*n*) Quite a number of the men resented the new regulation, but a few didn't seem to mind.
(*o*) Sound will only travel in air.
(*p*) Heat usually travels by conduction or a similar method.
(*q*) The reciprocating parts of a steam turbine set up a strain on the bearings.
(*r*) A glider pilot climbs on columns of hot air.
(*s*) When you put the car brakes on they have a sort of effect about the centre of gravity.
(*t*) Torque is a kind of effect of twisting motion.

B.

(*a*) Nuclear fall-out cannot be that dangerous. We have had no real signs of genetic upset.
(*b*) A coulomb is an excess of positive electricity on a body.
(*c*) Five out of a hundred parts, roughly 6 per cent, were defective.
(*d*) Aircraft are out of date; now it is all a question of rockets.

(*e*) The idea of the transistor is a completely new electronic discovery.
(*f*) Most workmen are careless, ignoring safety notices and becoming too cocksure of their own abilities.
(*g*) There is not much difference between the tractive resistance of locomotive tyres and rails and of car tyres and road surface.
(*h*) A refrigerator has to get hot before it gets cold.
(*i*) The image in our eye is not upside down like the one caused by the camera lens.
(*j*) Galileo showed that gravity pulled everything down the same way.
(*k*) Mechanics shows how things move about in space and how long it takes them.
(*l*) This film camera has two speeds; with one you can make things go more quickly, the other slows things down a bit.
(*m*) The differential in the back axle of a motor car is to help it to go round corners more easily.
(*n*) To be a skilled workman means to most ordinary people one who is able to do a variety of jobs on a lathe or similar machines.
(*o*) You should regard safety as important as any consideration in the workshop.

BE CONSISTENT Deal with one topic at a time

Do not: shift the subject in mid-sentence
cram different ideas into one sentence
muddle classifications—*e.g.*, saying that "a lathe is a process. . . ." when it is a *machine*
mix metaphors

e.g. Although a turner appears to be only a machine minder, turning is a skilled occupation.
The subject has shifted from *turner* to *turning*.
Alter the sentence:
Although a turner appears to be only a machine minder, he really follows a skilled occupation.
The foreman did not beat about the bush but put his cards clearly on the table.
This is an example of a mixed metaphor—*beating about the bush* and *playing cards* are used metaphorically but clash in a ludicrous way.

Rid the sentence of one metaphor:

The foreman did not beat about the bush but stated directly and clearly what he meant.

or both:

The foreman did not hesitate or digress but stated directly and clearly what he meant.

Exercises

Each of the following sentences contains some inconsistency. Explain what this inconsistency is and rewrite the sentence in an improved form.

A.

(*a*) Although drilling is an ancient technique the modern machines have altered out of all recognition.

(*b*) The pressure on the beam was so great that it eventually cracked and the strain gauges read correctly.

(*c*) The man who operates the lathe is an important cog in the field of industry.

(*d*) Reading a dictionary is a good practice, it gives you the words clearly shown, with help in the pro̅nunciation, and it also gives the meaning and derivation of the word.

(*e*) By measuring at one yard intervals along the metal strip, the holes for the bolts are located.

(*f*) Chatter is when something is loose in the tool holder and causes uneven finish.

(*g*) The casting was defective but he was not a very good moulder.

(*h*) Logarithms mean adding-up and subtracting instead of having to do long multiplying and dividing.

(*i*) If the men in this workshop do not pull their socks up they will find themselves in hot water.

(*j*) In processing a film you must first place it in a developing fluid, then you must transfer it to a fixer and secondly you must wash the film thoroughly.

(*k*) If a boy does not like getting his hands dirty or cannot put up with a lot of noise you will find a mechanic's career an unsuitable one.

(*l*) The evolution of the aeroplane was from gliders and kites, and in fact it looked in its first form like a box kite.

(*m*) Modern machine tools are much more accurate than the methods of turning in the early days of the industrial revolution.

(*n*) The shortage of natural materials has largely been replaced by the new synthetic compositions.

(*o*) The oil should be drained when the engine is hot, then fill with flushing oil after which you should let the engine tick over then drain the flushing oil before you put in the fresh oil.

B.

(*a*) A theodolite is a method of measuring horizontal and vertical angles.

(*b*) If you want to get equilibrium in a body, it has to have its c of g vertically lying over the base you support it with.

(*c*) Á centrifugal pump means having fans turning to push out the liquid.

(*d*) Grinding is useful not only for a lot of rough work, but you can use a grinding machine for finer and more accurate work just as well.

(*e*) Ohms law is when a current is at a constant temperature in an electrical circuit, and is proportional to the pressure on its terminals.

(*f*) If we want to get the Mach number of an aircraft in flight, we employ the method known as a machmeter.

(*g*) A megawatt is what we call a million watts.

(*h*) In loading the camera the cassette must be in the right hand, then feed the film into the take-up spool.

(*i*) The sports club is almost on the rocks, and if it is to play its part in the sphere of social activity it must have a good shot in the arm.

(*j*) The work in the research station centres around the biochemistry section.

(*k*) Insulin is the accepted treatment for diabetes, it is a disease which has been overcome by modern methods of analysis.

(*l*) The magneto is a form of induction coil used to spark the mixture of gases required by an internal combustion engine, it is somewhat old-fashioned; but another form of induction coil is the telephone.

(*m*) There was a flood of complaints from the men on piece work but they soon began to realize that they were banging their heads against a brick wall.

(*n*) Periodic is when something moves the same way after a certain time has passed.

(*o*) Dynamics is something to do with moving bodies, but when we deal with the motion of bodies when we do not have to consider the forces leading to the motion, we call that kinematics.

BE DIRECT Write sentences economically, and as briefly as possible.

Do not: Use padding
repeat information unnecessarily
use unnecessary abstract nouns
use wasteful introductory words
make irrelevant side comments
use pompous words and phrases

e.g. For all intents and purposes we may describe the *mol* as follows: A mol of gas is to be considered as that quantity of gas which possesses a weight which is equal to the molecular weight of that gas expressed in pounds.

This is too long-winded and could be rephrased:

A *mol* of gas is the quantity of gas the weight of which is equal to the molecular weight in pounds.

EXERCISES

Rewrite each of the following sentences in a more economical form. Rid the sentence of wasted words and phrases; look for repetition and pompous vocabulary.

A.

(*a*) It is essential that this machinery must be serviced regularly.
(*b*) The work piece was slowly drilled until the drill had drilled to a depth of about 22 mm.
(*c*) If you run out of paint and want to get a further supply the same as the first you will have to be careful or you will find that the second colour will not always match with the first, but there will perhaps be a very slight difference in the shade.
(*d*) Whatever the type of car it is useful to have a temperature gauge for the water system to show the driver the temperature of his engine and to warn him if there is any overheating, due to low water level or to a choked radiator, likely to lead to a dangerous rise in temperature.
(*e*) A selected few of the operatives were made the recipients of a special award for the longevity of their service to the company and the time they had been associated with it.
(*f*) He was unable to repay in full the debt of five pounds which he owed to his friend.
(*g*) Without doubt we are in need of a change of policy with regard to the care and welfare of staff.

(*h*) The purpose of this attachment is to see that the tool is lifted clear of the work each time it returns at the end of a cutting stroke.

(*i*) The meeting of the local branch of the union was making every earnest endeavour to bring a sense of urgency to the question of whether or not united action should be taken in view of the threat of certain men being no longer needed in the workshop.

(*j*) It is, as you know, a point of argument that heavy goods should be transported by railway and not by other means of transport.

(*k*) Modern aircraft, or at least those of a certain size and power, for example, jet air liners, need special runways, often quite long, before they can be operated successfully and with some sense of safety and security.

(*l*) The texture of the ground is good and strong and excellent for a solid foundation to withstand a building of 33 m if necessary.

(*m*) 'Racing' is, as you will know if you have read anything about engineering, the term we give to the rise in speed of an electric motor to a level which is agreed to be a dangerous one; it is also a term used in other branches of engineering, for example to describe a similar state of affairs in a motor car engine.

(*n*) The use of a convector heater is a means of providing economical space heating by operating at a low temperature whilst spreading out the heat produced to warm the surrounding air.

(*o*) There is a cell, called the photo-electric cell, which is characterized by its ability to change its electrical characteristics when light falls upon it.

B.

(*a*) A helical gear is a gear which has teeth but which has the wheel teeth not at right angles to the other teeth, but at some other angle, and is really a modification of the old stepped gear.

(*b*) It was desired for a report to be made out on whether it is an economical and worthwhile proposition to modernize completely the machine shop of our works at Crossley.

(*c*) A derrick is a form of lifting device which can be operated by hand by a single individual or which can be operated by means of a number of motors.

(*d*) There is a lever located on the apron of a lathe by means of which it is possible for the driving motor to be controlled through the shaft, enabling the motor to be stopped, started, or reversed.

(*e*) The firm was initially contacted with regard to initial estimates and a statement was made that they were not aware that the estimates were being asked for.

(*f*) As I have said before, there is quite a lot of unemployment in the area of this town and unless we are prepared as industrialists to open new branches of industry locally there will be much hardship in this area.

(*g*) Owing to the wasteful character of the process, it was decided by the management that they would institute a work study project to try to find a way to decrease the waste and consequently to make a considerable improvement in the upward rate of production.

(*h*) It was investigated as to whether a new factory was necessary in view of the fact that there was a possibility of extending the existing premises as they stand within the course of a few years.

(*i*) In all objects subject to the force of gravity there is a point at which that subject can be supported in a state of equilibrium and this is the point at which the force of gravity seems to be applied.

(*j*) It is advisable that all apprentices in this organization should benefit from part-time education and that they should not only attend the local technical college for one day a week but should also participate in the apprentice training scheme devised by the firm.

BE METHODICAL Arrange words carefully to give a clear direct impression

Do not: twist the word order
add afterthoughts
misplace phrases, causing ambiguity
use overlong sentences
depend too much on *and* and *but*

e.g. Duralumin is a very useful alloy of aluminium, for it can be used for forgings and rolled sections, but it is not any use for castings, though it is used for pressings.

This is a clumsily constructed sentence with an awkward afterthought. Compare the following version:

Duralumin, an aluminium alloy, is used for forgings, pressings and rolled sections. It is not suitable for castings.

EXERCISES

The following sentences are examples of clumsy arrangement. Rewrite them in a clearer and more methodical form.

A.

(*a*) The workers will have less eyestrain due to the better lighting.

(*b*) The reading can be set on the dial by either turning it clockwise or anticlockwise.

(*c*) This brand of lubricating oil is as good as, if not better, than the one we have been using.

(*d*) This work needs doing badly or it will be too late.

(*e*) A new bonus and incentive scheme was introduced into the workshop which was not popular.

(*f*) They managed to equip the new pavilion from the proceeds of both a dance, and a series of social events, and money-raising activities, also three jumble sales.

(*g*) You can make a Pelton wheel, which is a powerful little motor that is driven by water, by shaping a disc of wood, and marking pieces of tin which you fix into the wood and then fit the wheel to a spindle and mount the wheel in a case which has a pipe to direct jets of water against the blades, and the wheels will revolve at high speed.

(*h*) He placed the piece of metal in the vice then hammered it.

(*i*) There was a breakdown in the process so the machine parts on the assembly line had to be removed quickly and disposed of before they piled up and became a serious problem.

(*j*) Sand castings usually have pieces of extra metal left on them and these have to be trimmed and cleaned off and this we call 'fettling.'

B.

(*a*) Next after the stern gland we come to the business end of the craft, that supplier of driving force—the propeller.

(*b*) This is not only an example of human fatigue problems but problems of human adjustment to the environment.

(*c*) In the top section of the rocket was a compressed air compartment with a diameter of 1 metre which worked the control gear.

(*d*) An immersion heater consists of an electrical equipment immersed in whatever you want to heat usually with a thermostat or by hand.

(*e*) I am so pleased with the car, so is my wife. Of course we are. It comes from . . . Hundreds of satisfied customers, and dissatisfied, both of whom have since discovered how wrong they are.

(*f*) Friction can be reduced between two solid bodies by means of either changing the motion to a rolling motion by ball bearings or roller bearings, or by introduction of a lubricant between the contact surfaces.

(*g*) When light or sound waves come to someone from a moving place of origin, then those waves are somehow squashed and will consequently appear to the observer to be shorter, and waves going away provide the opposite effect, which is known as the Doppler effect.

(*h*) Heating by conduction is not the only method, and if you could only heat water by conduction it would take a long time, but because there is heating by convection you are able to heat water first by contact with the heat source—conduction—and then by movement of the heated water—convection.

(*i*) Modern telescopes fall into two categories—refracting and reflecting, but the reflector telescopes are the more common, mainly because the refractor has four surfaces to its achromatic objective, whereas the reflector has but one, and the grinding of a perfect surface is a long process.

(*j*) A cam is a body which gives motion to another body which is called a follower and with which it is in contact and which is, like itself, reciprocating, rotating or oscillating, and the motion is in accordance with the shape of the cam and of the contact surface of the follower.

BE INTERESTING Style is important. (Think before you begin to write.) Write in a smooth easy manner, varying the length and structure of the sentences.

Do not: write sentences that *sound* awkward and jerky

string together sentences of similar construction

e.g. Atoms are too small to be seen by the most powerful microscopes. They are not solid but almost all space. They consist of a central charge of electricity. This charge is positive. It is called the proton. Around the proton negative charges, or electrons travel at great speed. The movement is somelike that of planets around the sun.

The sentences are too short and jerky. Some of them can be combined to give a smoother effect.

Atoms, which are too small to be seen by the most powerful microscopes, are not solid but almost all space. They consist of a central positive charge of electricity called the proton. Around this proton negative charges, called electrons, travel at great speed, like planets whirling about the sun.

N.B. Read your work 'silently aloud' to make sure that you have the right harmony and rhythm.

EXERCISES

A. Convert each of the following series of brief statements into one interesting and well written sentence. You may alter the form but not the meaning of the statements, and may use them in any order.

(*a*) Domestic fires are wasteful.
Much heat goes up the chimney.
Smoke from fires causes smog.

(*b*) Carbon is a black, uninteresting solid.
It has beautiful crystals.
The rough crystals are called diamonds.
They reflect light in rainbow hues.

(*c*) Radar has many uses.
Radar can be used in war or peace.
It can detect enemy ships and aircraft.
Collisions in fog can be avoided.
Control at airports is made easier.

(*d*) A thermostat can maintain a specific temperature.
Thermostats can be made from bi-metal strips.
Control can be obtained by cutting current in and off.

(*e*) Rocket propulsion depends upon action and reaction.
A rocket will work in a vacuum.
Space ships depend largely upon rocket power.

(*f*) An explosion liberates energy.
This energy is liberated in large quantities.
This is used in the internal combustion engine.
A mixture of gas and air is exploded in an iron cylinder.

(*g*) Centuries ago people used an abacus.
Counting to-day can be done more quickly.
Modern computers work faster than man.
Electronic valves do the work of brain cells.

(*h*) Cumulus clouds are common during hot summers.
Moist air rises and water vapour condenses.
Water droplets form a cloud.

(*i*) The ramjet depends upon air entering at high speed.
This air helps the fuel to burn in the burning chamber.
Hot burning products rush out at the rear and push the jet forward.
The ramjet is sometimes known as the 'flying smokestack.'

(*j*) Special motor roads are safe.
High speeds need certain precautions.
Vehicles behave differently at constant high speeds.
Tyre pressures need to be higher under such conditions.

(*k*) Fibre glass is used for car bodies and for boats.
It has a high tensile strength.
Fibre glass needs only cleaning by detergent.
Repairs to it are simple and straightforward.

(*l*) Safety measures are important.
There are many deaths in the home and the factory.
Some accidents are caused by carelessness.
Workmen can become overconfident.
Machines and kitchen equipment must have safety guards.
People must be constantly reminded.

(*m*) Meteorites entering the earth's atmosphere begin to burn up.
Space travel involves the problem of re-entry.
The first astronauts accomplished this successfully.
The outer skin of their capsule was still very hot.

(*n*) Transistors have transformed the radio industry.
Miniature radio and television are a reality.
There is no loss of volume.
Transistors have made possible new electronic machinery.

(*o*) Modern science helps to combat crime.
Ballistics helps to identify bullets.
Infra red techniques identify forgeries.
A speck of blood can be traced by its group.
The microscope is a valuable police instrument.

B. In each of the following extracts the sentences are short and jerky. Link them sensibly, rephrasing where necessary, to produce a smoother passage.

(*a*) Plants are provided with roots. These help them to stay upright in the ground and to obtain food from the soil. Plants need nitrogen. Nitrogen is in the air. The plant can only take in nitrogen when it is in solution. This means that rain is important. The plant takes in the solution by capillary attraction.

(*b*) Lime is made by heating limestone. It used to be the commonly used mortar for builders. It was good for ceiling and wall plasters but it was slow to harden. This was because it had to take carbon dioxide from the air. This then turned it back to hard limestone. Cement has now taken over from lime. It is quick setting.

(*c*) A brake brings to rest a body in motion. It is also used to hold a body in a state of rest. It can be used as well to keep a body in a state of rest or motion against the action of external forces.

(*d*) Craft apprentices are expected to know some arithmetic. They should be interested in science and handicrafts. They should be prepared to face a five-year course of study. Their initial training will be in the apprentice workshop. They will also be attending day-release classes. This will be for City and Guilds examinations.

(*e*) Surveying involves measurement. Often you need to mark lines at right angles to each other. If you need to be accurate a theodolite must be used. If only approximate readings are required a cross staff can be used. This is especially the case when taking offsets.

(*f*) Gas turbines are used in some aircraft. They drive the propellers and consist of an axial-flow compressor. This is driven by a two-stage turbine on the same shaft. Around the engine are combustion chambers. The propeller is driven by a separate turbine wheel. There is a jet pipe at the rear. Thrust is obtained from ejecting hot gases.

(*g*) The fluorescent mercury-vapour lamp has a tube containing a mixture of argon and mercury. The tube is glass and is coated inside with some fluorescent powder. Tungsten electrodes are at each end. Each electrode has a nickel strip as an anode. A thermal switch is placed in a glass bulb filled with hydrogen. This is used to start the lamp.

(*h*) Colour vision depends on light sensitive elements in the retina. There are only three of these elements. Each of them responds to a broad band of wave lengths. Low frequency monochromatic light excites one of the elements and produces what we call 'red' sensation of colour. The medium frequency acts similarly. So does the high frequency. They produce respectively sensations of green and blue. Intermediate frequencies excite two sensitive elements. This produces a mixed colour. For example red and green elements together produce the sensation of yellow.

(*i*) An important law is the second law of thermodynamics.

This states that heat cannot pass spontaneously from a colder to a hotter body. This means that a refrigerator needs an outside source of energy. This energy must produce a temperature difference. This is achieved by using a liquid with a boiling point below the freezing point of water. This can provide a cooling cycle. The liquid is evaporated at a low temperature. It absorbs heat from inside the refrigerator. It is then compressed, usually by an electric motor and heated. It passes through a condenser. It becomes a liquid once more and the cycle is resumed.

General Exercises

The following short passages are actual examples of students' work. Study them carefully, then rephrase each passage in good English.

1. From descriptions of a lathe.

(*a*) In the feedshaft is contained the traverse up and down the guideways on the lathe bed and the traverse across the bed.

(*b*) Below the chuck can be seen a gap and this allows a larger swing of the workpiece.

(*c*) The bed is made of cast iron and forms as a support of the lathe on top of it is guide ways, which the saddle and tailstock are fixed to and enables them to travel along the lathe.

(*d*) The chuck is rotated by a transmission system of pulleys or gears used driven by an electric motor, the range of gears used on a lathe depends entirely upon the work required and the range of work sizes.

(*e*) On the front of the headstock is a chuck, that is a cast piece of iron that is one large clamp. This holds the workpiece and is revolved at high speeds by means of changing the gears.

(*f*) The clutch is normally two circular sheets of metal with cork on one side. The cork engages with each other when the handle is moved so engaging the motor and main spindle.

(*g*) Surmounted on the saddle is a toolpost in which the cutter's holder is held, this part may be swivelled round and allow the cutting tool to work at an angle to the work piece.

(*h*) The headstock is made of cast iron and within it are the motor for driving the lathe which is connected by vee belts to the gears of the lathe which rotate the chuck, which faces the rest of the lathe, and which drive the feed shafts, which are under the lathe bed; all of these gears being horizontally above the motor in the headstock.

2. From a description of a hacksaw. (With this example it would be best to start again.)

The frame of the saw is in the shape of a 3-sided rectangle. The fourth side, and equal to the opposite longer side, is the blade which can be removed.

3. From a description of a sparking-plug.

A simple form of sparking-plug will consist of three main parts. Its most simple outlook will be of three cylinderical solid objects, of three different diameters all standing on each other. The largest diameter metal cylinderical casing, standing at the botton, fixed into this casing but projecting above is a poclein bit. This poclein bit acts as the insulator. Running down the middle of the poclein insulator is a metal rod which project above the poclein insulator, forming the topmost cylinderical looking object in the basic view—the poclein insulator being the middle cylinderical looking object.

PUNCTUATION

Do not be too preoccupied with the *rules* of punctuation. Be conscious of the *purposes:*

To help to make the meaning clear
To emphasize particular words and phrases
To mark stages in the development of ideas
To help the reader to move quickly through the material.

Punctuation marks are instructions to your reader. Look at your writing from *his* point of view, and use those punctuation marks which will be most helpful to him.

The following suggestions are for your guidance in using particular punctuation marks to help the reader:

Full Stop

This mark of punctuation tells the reader that he has come to the end of a particular statement. He knows that the thought you have expressed will be followed by another, or by a related idea.

Full stops in a paragraph indicate the number of ideas that the reader will have to deal with. They are also an indication of the concentration he requires—too long a sentence can mean an inattentive reader. Do not, therefore, spend too long between full stops.

Semi-colon

The semi-colon tells the reader that he has read one aspect of the idea you are expressing, and he is now to hear a contrasting idea, or another version.

e.g. Cold working of steel is carried out at the pearlite stage; hot working is carried out at above the lower critical temperature.

Comma

The comma is the most frequent mark of punctuation. Just as the full-stop tells the reader he has come to the end of a statement, so the comma indicates to the reader the various sections of that statement. The comma is a guide to the grammatical construction of your writing, and prevents the reader from misunderstanding you.

Commas enable the reader to:

recognize words in a list.
sort the main ideas from the subordinate ones.
distinguish the main flow of ideas from asides and additional comment.
maintain concentration by dividing a long sentence into manageable sections.

The following sentences contain examples of the use of the comma. Study each one carefully and discover why the commas are useful.

Fluorine, a greeny-yellow gas, is more poisonous than chlorine.

The intermediate gear has no effect on the ratio between driver and driven, but acts as a connexion.

The works manager, who used to be with a large export firm, has distinctly improved the factory organization.

John, having completed the job, packed his tools away.

Colon

The colon is an introduction mark. It tells the reader that an explanation is to follow, or that an illustration will be given. Examine the following example:

Two problems are outstanding: the acquisition of suitable property and the engaging of skilled staff.

Dashes

A pair of dashes usually indicates to the reader that there is an interruption of the flow of thought, perhaps to express another idea or to make a side comment.

e.g. The strict observance of safety rules—a matter of some importance in this organization—was the subject of last week's bulletin.

Study punctuation in your reading. Notice how bad punctuation can hinder you as a reader and determine that you will always show consideration for *your* readers. Always edit your work with an eye to the most suitable punctuation.

Exercises

The best way to practise punctuation is to write and then submit your work for criticism by your lecturer. The following exercises are provided for some practice in criticism and in the specific use of marks of punctuation.

1. Criticize the punctuation used in the following sentences:
   (*a*) The temperature was not high enough, the two metals would not fuse.
   (*b*) Three hours, four minutes were allowed for the journey, however that time was not needed.
   (*c*) The lecturer said, that the hammer is the most used of all the fitters' tools.
   (*d*) Perhaps, careful observation for a few days, will give us fresh information.
   (*e*) An engineer, who wishes to climb to the top of his profession, must prove his worth.
   (*f*) The plate, having been cleaned thoroughly, you should smear it with a layer of oil.
   (*g*) Some atoms are unstable but they do not break down all at once.
   (*h*) Below some workman were signalling for the crane jib, to move further towards the excavation.
   (*i*) The machine is now obsolete, and can be dispensed with.
   (*j*) The solid, fluted, machine reamer differs from the hand reamer, in having a taper shank.

(*k*) A self-centring chuck can be used for light-turning jobs.
(*l*) The resultant diagram figure 32, shows little deviation from the normal.
(*m*) He knew the oil was low in viscosity; but he did not know what other lubricant to use?

2. Make each of the following sentences or passages clearer by a judicious use of punctuation:

A.

(*a*) The equipment for a lathe includes tapers reamers steadies and chucks.
(*b*) The material will be delivered on Thursday 5th January 1961.
(*c*) The job having been finished the men went home.
(*d*) The machine shop is the busiest section employing over 60 per cent of the staff.
(*e*) If the metal tears the tool cut is obviously too deep.
(*f*) He exposed the film with an aperture of f16 at 250th of a second developed it in a fine grain developer for the specified time and printed the results on a soft paper.
(*g*) There is one hope of overcoming the difficulty we can switch the work to other machines.
(*h*) Belt drives power presses milling machines lathes lifting tackle all these come under the watchful eyes of the factory inspectors.
(*i*) Assemble the apparatus Diagram 1 fill the glass tube with mercury expelling all air invert the tube under the surface of the mercury in the dish.
(*j*) A pile is a timber steel or reinforced concrete post usually vertical and less than 60 cm square driven jacked or jetted into or cast into the ground.
(*k*) After the rocket has reached its maximum velocity perhaps one kilometre a second the stage of fuel cut-off is reached called by the experts brennschluss end of burning period.
(*l*) Although the production of steel castings is in general similar to that of iron castings the higher melting point of steel makes founding more difficult and the extent of contraction when steel cools causes more cracking.
(*m*) A lead sheet width 1 m of indefinite length is bent to form an open gutter of rectangular section the area of the cross section is A sq cm see fig 7 what is A expressed as a function of the width of the gutter b.
(*n*) Wherever we go in home factory shop club school and in most forms of transport we find the growing use of synthetic

materials particularly the forms we have come to know as plastics polythene polyesters of all descriptions materials developed in the laboratory to meet modern demands.

(*o*) The Hovercraft which has been given the designation SRN 1 is powered by a 335.25 kW engine driving a ducted fan this supplies propulsion and lift so that the craft can operate at about seven feet above the water moving in fact upon a cushion of air generated between the hull and the surface of the water.

B.

(*a*) Our latest machine which has been installed in the new workshop is expected to give good service this it must be said despite some adverse criticism.

(*b*) The ordinary milling machine although it lacks most of the refinements of the more complicated models is of solid construction and serves a very important routine purpose in the workshop.

(*c*) Recently as a result of much intensive research there has been a great improvement in methods of photographic colour-printing but it is still not easy for the amateur to gain the correct colour balance so exacting are the requirements of temperature and of filter factors.

(*d*) Information about new automatic processes including the technique of electro-erosion is published in the monthly journal *Technical Advance* obtainable by annual subscription only from S.T.A. Society for Technical Advance Lowndes Square London W.1.

(*e*) Triangulation the application of trigonometry to surveying involves the measurement of large sections of land by dividing it into triangles of which (a) the base line is measured and (b) the angles are measured accurately so enabling all the other sides to be calculated.

(*f*) Many industrialists are short sighted when it comes to improved design of machinery even when the expense involved in re equipping or adapting is only nominal and a typical response to suggestions is the remark made to design consultants. We cannot risk interfering with production merely to add some gimmick or other.

(*g*) Fresnel a great French scientist began his working life as an engineer then became interested in optics especially lenses as a means of improving lighthouses. When he died at the early age of thirty-nine he had already made great contributions to science the annular lens the dioptric lens and mirrors replaced by prisms.

(*h*) The ear consists of three parts the inner middle and outer. Air travels through the cavity of the outer ear the vibrations being felt by the eardrum the middle ear with its three bones is connected to the throat and controls the air pressure the inner ear contains the nerves which respond to the vibrations.

(*i*) The behaviour of molecules in a gas is predictable and can be described mathematically molecular behaviour in liquids is not simple defying mathematical formulae with its random movement.

(*j*) A three stage space rocket consisting of first stage rocket motors second stage storage tanks and driving motors and third stage capsule must achieve a velocity of 11.2 km/s if it is to escape from the earth's gravity and move out into space. Even when this formidable barrier is overcome the space traveller still faces the difficulties of precise navigation and the most formidable problem of all re entry into the earth's atmosphere.

II

PARAGRAPHS

1. The Paragraph as a Unit

Sentences are used in groups to form the next larger unit of expression—the paragraph. There comes a time when the idea to be expressed is too long or complicated for it to be fitted within the framework of a sentence. Examine the following sentence dealing with the welding of wrought iron and mild steel:

> It is possible to weld wrought iron and mild steel by pressing together two surfaces of the metal at the right welding heat which is about 1·350° C. for iron, when it is white hot and in a pasty condition, and a little lower in temperature for mild steel, watching for a yellow colour before it merges into white; but the temperature must be correct because if it is too low it will not weld and if it is too high the metal will be burnt, and every effort must be made to ensure that the surfaces to be welded are scrupulously clean, this being effected by the use of a suitable flux.

The information has been crammed into a long sentence which has become too clumsy to be of any value. The sentence form is unsuitable and the information must be expressed in the larger paragraph unit. Look at the following example:

> It is possible to weld wrought iron and mild steel. This is done by pressing together two surfaces of the metal at the right welding heat. This is about 1350° C. for iron. Then it is white hot and in a pasty condition. A little lower temperature is needed for mild steel. With this, welding should take place when the metal is yellow before it merges into white. The temperature for welding must be correct. Too low a temperature and the weld will not take. Too high a temperature and the metal will be burnt. Every effort must be made to ensure the surfaces for welding are scrupulously clean. This can be effected by the use of a suitable flux.

The sentence has been broken up into a series of sentence units, but these sentences are short and monotonous, creating a jerky effect throughout the paragraph. A smoother effect is needed if the paragraph is to be easy to read. Sentence length and structure must be varied.

> It is possible to weld wrought iron and mild steel by pressing together two surfaces of the metal at the right welding heat. For iron the temperature required is 1350° C., when the metal is white hot and in a pasty condition. Welding takes place at a lower temperature with mild steel, when the metal is yellow and just before it merges into white. The correct temperature is important, because too low a temperature will prevent the metals joining, and too high a temperature will cause the metal to burn. The added precaution must be taken of keeping the welding surfaces scrupulously clean with the aid of a suitable flux.

Even this paragraph is by no means perfect, but the effect of jerkiness has gone.

EXERCISES

1. Draft several versions of the above paragraph, trying each time to improve the expression. Experiment with sentence arrangement, word order, and linking-words. Discuss in your group the comparative merits of different versions.
2. Study the following clumsy multiple sentences and recast them in paragraph form.

(*a*) Energy, which makes work possible, is used up every time we move our limbs, or every time the heart beats or the lungs expand, and this activity continues night and day, maintaining as an added task, the temperature we need to keep warm and alive, and expending in this operation enough equivalent power to bring a large measure of cold water to the boil.

(*b*) If we wish to discover facts we may use our eyes to observe the world around us or use them to search in books for facts and information which have been collected for our benefit for over thousands of years, for the world's knowledge is largely contained in the millions of books in its various libraries, ready for the eager research worker but not prepared to give up that information without some effort on his part, particularly the effort of studying the best ways of extracting that information from the bewildering mass of print.

(*c*) The growing demand for small, efficient motor cars has led to the development of simpler methods of lubrication, transmission, and structure, three factors which have an important bearing on the durability and life-span of the car—the lubrication is increasingly carried out by the owner, who does not wish to wriggle about with a grease gun to deal with twenty or more greasing points, but welcomes the new rubber-cone suspension units which obviate the need to lubricate, and also welcomes the single lubrication unit evolved in some of the front-wheel-drive cars where the sump serves the engine and transmission.

(*d*) Metal has many advantages over stone including its greater strength—which means that a metal support will carry much greater weight than a stone one of the same size—but many metals are very heavy and are more difficult to move and manage than stone, except aluminium which is only about a third of the weight of iron, but is stronger than stone, two factors which make it very useful in the manufacture of such items as motor cars, aeroplanes, and railway coaches, especially when aluminium is strengthened by the addition of other metals to make it into an alloy.

(*e*) Safety precautions are essential in a workshop, particularly where apprentices, who are naturally not 'machine-conscious,' are working with dangerous machines, such as milling machines or grinders, but the act of putting up warning notices is not sufficient, for the 'old hands' in the machine shop tend to become careless (familiarity breeds contempt!) and their attitude is communicated to the young people they are teaching, which means that the need for safety should be one of the most important propaganda considerations in a factory and the right attitude to it created in all possible ways.

2. The Structure of a Paragraph

(i) *The Topic Sentence*

Just as a sentence consists of a main idea built up into a context by modifying and qualifying words, so a paragraph requires a main idea which is built up into a larger context by modifying and qualifying sentences.

This main idea is expressed in a *Topic Sentence*. If you read once more the sample paragraph on page (38) (the third version) you will be able to identify the topic sentence quickly:

> It is possible to weld wrought iron and mild steel by pressing together two surfaces of the metal at the right welding heat.

The paragraph goes on to give further information about the conditions under which welding can take place.

EXERCISE

Read the following paragraphs and pick out the topic sentence in each:

(*a*) Hot rolling of steel produces sheet, strip, or bar steel with a non-polished reddish blue surface. The steel ingot from the ingot mould is almost at white heat and is reduced to blooms at the blooming mill. The blooms are then cut into lengths and pressed backwards and forward through the roll openings of the rolling mill or through the various operations of the continuous mill, until they attain the required form.

(*b*) The old-fashioned method of finding the depth of water beneath a ship's hull was by a weighted line. The simple lead line consisting of a lead weight on a length of rope was swung and dropped alongside the ship and the depth read off according to the various coloured markers fixed to the line. For readings in deeper water, a Kelvin deep-sea sounder, with a longer steel wire and heavy sinker, was wound out from a drum—an operation which might take several hours.

(*c*) Human beings have eyes in the front of the head for much the same reasons that predatory animals have them so placed. The binocular vision given by the frontal position of the eyes means an efficient method of estimating distance. This enables the hawk to descend unerringly on his prey and the cat to pounce swiftly and surely upon a mouse.

(*d*) As early as in Egyptian times, special 'chainmen' measured the cornfields along the fertile banks of the Nile. Roman officials measured land and roads all over their extensive domain. Visitations of Norman officials collected entries for the famous 'Domesday Book.' Even in early times the development of society was linked to the consideration of land measurement and ownership.

(*e*) During the Second World War there was a demand for war chemicals—explosives, lubricants, fuels of all kinds—and a need for many materials that were in short supply or completely unobtainable. Rubber, alcohol, and cleaning fluids had to be replaced by synthetics, and from this research and experiment came a new chemical product. The production of plastics and detergents and the development of our huge synthetics industry owe much to wartime research.

Identifying the topic sentence should have been easy in those examples; but many writers are careless in their paragraph construction and expect you to rummage through the sentences to find the main idea. Make certain that you are one of the *careful* writers.

(ii) *Developing the Topic Sentence*

The Topic sentence may be expanded in a number of ways, or it may come at the climax of a series of preparatory sentences. The following are some simple examples of paragraph development. The Topic sentences are in italics.

(*a*) Giving examples of the statement made in the topic sentence:

A workman in a machine shop can be in danger of injury at any time. Loose clothing can catch in the working parts of a machine or in a belt drive that is not carefully screened. Heavy, loose tools may be dislodged and fall against him, or he may lose his balance as the result of striking a low projection or slipping on a greasy part of the floor. Heated or electrically charged tools may slip and cause serious burns if his hands and body are unprotected.

(*b*) Extending a comparison or contrast made in the topic sentence:

Even in a basic operation such as filing, there is an obvious difference between the skilled and unskilled worker. The young apprentice will press too hard with his right hand at the beginning of the stroke, and will shift to a further hard pressure from the other hand as he finishes the stroke. The skilled man will keep the file horizontal for the whole stroke, obtaining a flat even surface.

(*c*) Explanation of a statement made in the topic sentence:

Tin cry is a useful aid in judging the quality of a solder. If a metal containing tin is bent near the ear, you will hear a crinkling sound which increases with the proportion of tin present. It is caused by a deformation of crystals, and a solder that produces a loud tin cry is usually one that contains a large amount of tin.

(*d*) Providing reasons for, or proof of a statement made in the topic sentence; this means that the topic sentence may come *after* the sentences giving the proof:

The atmosphere of the planet Mars contains oxygen, the continued existence of which suggests a system of plant life to free the oxygen after combination with the planet's crust. The surface of Mars also shows signs of seasonal change, suggesting the annual growth and development of vegetation; now, during a recent dust storm on Mars, scientists believe that the gradual reappearance of patches of vegetation through the haze indicates the ability of such vegetation to throw off the dust, and such activity may not be confined to plants. *There is reason to believe that life may exist on Mars.*

(*e*) Explaining a grouping or classification made in the topic sentence:

Trans-polar air travel is aided by four important instruments. Radar is used not to navigate by map, but to reckon ground speed by bouncing echoes off large objects shown on the screen. A directional gyro replaces the magnetic compass and may be attached to the automatic pilot. The best method of navigating is, however, by the sun and stars. In the summer months a sextant can be used, but during the long twilight, a special periscope polarized sky compass is needed to get a bearing of the sun even when it is hidden by cloud or beneath the horizon.

(*f*) Analogy, or the use of figurative language:

The nervous system resembles a complex telephone network. Messages arrive at the brain through the nerve wires, routed through the synapses; are recorded or passed on to the muscles and glands for action. Two main exchanges, the brain and the nerve centre in the spinal cord, deal with most messages and are linked by their junction at the uppermost part of the spine.

Those six paragraphs indicate a few ways in which a topic sentence may be developed into a paragraph unit. It is not, however, essential for the topic sentence to introduce the paragraph—sometimes the first sentence will link the idea with the previous paragraph before the topic sentence is introduced; sometimes the topic sentence will come as a concluding, and summing-up sentence.

Your choice of development is usually decided by the function of the paragraph and its relationship with those coming before and after it.

3. The Function of a Paragraph

A paragraph is a unit in a larger set of ideas—perhaps a section or chapter—when it may be:

(*a*) *a formal introduction to the ideas expressed in following paragraphs*
(*b*) *the development of part of an idea extending over several paragraphs*
(*c*) *a unit containing the development of a single idea*
(*d*) *a means of transition from one idea to another*
(*e*) *a summary of the ideas contained in the preceding paragraphs*
(*f*) *a formal conclusion to the set of ideas.*

It may also be a unit of direct speech.

Paragraph indention is a form of punctuation. It signifies that one paragraph has ended and that there is to be a slight pause before the next unit of expression begins.

The paragraph is also a *unit of concentration*. In considering your reader you must judge his ability to maintain concentration over a certain number of words or lines. Long paragraphs —particularly if they contain long words and sentences— stretch the reader's concentration, and if he has to go back over the paragraph to discover what it was about, you have wasted valuable time.

The length of a paragraph is largely determined by its purpose, and by the length of the work of which it is a part. Short paragraphs may be suitable for a one-page cyclostyled report, but they would look silly in a much larger work.

Transition paragraphs are usually very short, as are paragraphs involving surface comparisons and contrast. Those dealing with illustration and explanation tend to be long.

Variation of paragraph length in a report is as important as variation of sentence length within the paragraph. It creates a sense of rhythm and movement which aids the reading.

Examples of Paragraph Function

(*a*) *A formal introduction*

The following tests were carried out by our research team, and are presented here in correct sequence with a summary of the results of each test.

(Short and to the point.)

(*b*) *A section of a larger unit*

The second consideration must be to keep the surfaces clean. By means of a special flux it is possible to dissolve the scale and ash into a liquid slag when the flux melts. The flux then acts as a protective covering to the surfaces. Borax could be used for mild steel, while sand would be a good flux for wrought iron.

(Part of a wider subject, *welding*, and of a section dealing with precautions to be taken.)

(*c*) *The development of a single idea*

Safety precautions are an important factor in industry. They ensure that time and money are not lost through direct injury or through the infection that might arise from a seemingly trivial cut or abrasion. They avoid the tragedy that comes with death or permanent disablement, freeing both employer and employee from anxiety. They foster, in consequence, some measure of confidence and good morale in the organization.

(Deals with the single idea of safety precautions as an important factor in industry. This does not mean that the paragraph necessarily stands by itself. It may be a unit within a discussion of industrial relations.)

(*d*) *Transition*

Having considered which types of lathe to install in workshop B, we must decide the number and type of milling machines required in the adjoining shop.

(The paragraph indicates the change from 'lathes' to 'milling machines.')

(*e*) *Summary paragraph*

The investigation has, then, revealed three flaws in the process of manufacture. The siting of the machinery is unsuitable; insufficient time is allowed for the drilling operations; and the inspection system causes considerable delay.

(This paragraph refers back to the three main paragraphs or sets of paragraphs dealing with the three flaws.)

(*f*) *Concluding paragraph*

Perhaps, with these new methods, a mass-produced receiver will be possible. If so, coloured television will no longer be a purely technical process for special purposes, but will be an economic proposition for transmitting entertainment to the home.

EXERCISES

In the following exercises you will be asked to develop topic sentences into paragraphs. Some of the topics will be familiar to you, others will involve some research, probably in the library. Please remember that these are important exercises in the craftsmanship of writing—try not only to write correct English but to be interesting.

1. Develop each of the following topic sentences into a paragraph using the method of *explanation*:

A.

(*a*) In drilling it is important that all details should be correct.
(*b*) Loose clothing is dangerous in the machine shop.
(*c*) Filing is an individual and variable operation.
(*d*) Cleanliness is essential for successful soldering.
(*e*) The lines of force of a magnet can be revealed by iron filings.
(*f*) Burning requires air in addition to fuel.
(*g*) Effective earthing lessens the risk of dangerous shock.
(*h*) A photographic dark room should be free of dust.
(*i*) A simple try-square is one of the most useful of measuring tools.
(*j*) Heat from the sun travels by radiation.

B.

(*a*) Annealing is a process of heating and allowing the steel to cool very gradually.
(*b*) Jigs enable the repetition of certain drilling operations to be carried out speedily and accurately.
(*c*) To be effective a towel has to be made of a specific material.
(*d*) Artificial satellites are important for the solution of several problems.
(*e*) Controlled explosives are a useful source of energy.
(*f*) Penicillin and streptomycin are useful antibiotics.
(*g*) The calculus deals with quantities that are growing or diminishing.
(*h*) Differences between substances are the result of differences in their internal structures.
(*i*) Jet propulsion depends upon the recoil principle.
(*j*) Corrosion is a serious economic problem.

2. Develop each of the following topic sentences by *classification*:

A.

(*a*) There are three main adjustments to be made when setting a camera.
(*b*) Electric clocks are of two main types.
(*c*) Chatter on a shaping machine may be caused in several ways.
(*d*) There are three principal types of deformation of a solid body.
(*e*) The heat treatment of metal is to give it certain necessary properties.
(*f*) Taper turning on the lathe can be carried out in three ways.
(*g*) Science depends on observation, experiment, and measurement.
(*h*) There are two important conditions for an efficient bearing combination.
(*i*) Jet propulsion has a number of advantages.
(*j*) There are two main kinds of lever.

B.

(*a*) There are three essentials for the working of a heat engine.
(*b*) Hot working of steel includes several processes.
(*c*) There are several different forms of toothed gearing for transmission of motion or power.
(*d*) A glider pilot has several ways of obtaining lift for his aircraft.
(*e*) All quantities are classified into two classes—constants and variables.
(*f*) There are three qualities essential for the really successful engineer.
(*g*) A report usually consists of four sections.
(*h*) The traffic engineer can use several different methods for preventing chaos at traffic intersections.
(*i*) Servomechanisms are used for a number of important processes.
(*j*) Technical education provides courses for several different types of student.

3. Develop the following topic sentences by *contrast:*

A.

(*a*) Heat and temperature are two different things.
(*b*) Brazing ensures a tougher joint than one made by soldering.
(*c*) People in a workshop often have different attitudes from those of the office workers.
(*d*) Electric current may be transmitted in *A.C.* or *D.C.* form.
(*e*) Wire is drawn by either the 'wet' or 'dry' process.

(*f*) Hot and cold chisels differ in their cutting edges.
(*g*) The small firm is sociable but the large firm pays.
(*h*) Heating by electricity is more efficient than heating by coal or coke.
(*i*) Individual drive for machinery has advantages over the belt drive method.
(*j*) Some people prefer having spare time to working overtime.

B.

(*a*) The combination square is less generally effective and accurate than the simple try-square.
(*b*) A land surveyor has a more scientific job than that of a general surveyor.
(*c*) Anodised aluminium is more suitable than chromium plated metals for the finishing of car bodies.
(*d*) Thermostatic devices vary widely in the method of actuation.
(*e*) An electronic brain has not the portability of a human one.
(*f*) Transistors have several advantages over the old-fashioned valves.
(*g*) In boundary friction the oiliness of the lubricant is more important than its viscosity.
(*h*) The modern racing car has a greater overall efficiency than had the 'monsters' of the old tracks.
(*i*) Chemistry and biology differ in emphasis rather than in their fields of study.
(*j*) What a person means is often not what he says.

4. Develop the following topic sentences by *comparison*:

A.

(*a*) The lathe and the milling machine are both cutting tools.
(*b*) The first motor cars were very similar in appearance to the horse-drawn carriages they replaced.
(*c*) Sailing and gliding have something in common.
(*d*) You should aim and shoot with a camera as carefully as you would with a gun.
(*e*) The Goliath crane is like a large bridge.
(*f*) Dealing with radioactive materials is as tricky as handling explosives.
(*g*) A slide rule is as essential as a reference book.
(*h*) An engineer should be able to read a drawing as quickly and accurately as he can read a book.
(*i*) Sand can act like a liquid.
(*j*) Poor spelling is almost a national disease.

B.

(*a*) A bat moves by means of a system similar to that of radar.
(*b*) The microstructure of pure iron is like crazy paving.
(*c*) Fibre glass has many of the qualities of steel.
(*d*) The modern transistor is similar to the old 'cat's whisker.'
(*e*) A rocket is moved in the same way that a gun recoils or a boat moves when you jump ashore.
(*f*) An atom is constructed almost like a small planetary system.
(*g*) A cyclone can be described as an atmospheric whirlpool.
(*h*) Compared with differentiation, integration might be regarded as a 'return journey.'
(*i*) Thermionic emission is similar to the evaporation of liquids.
(*j*) The modern computer is based on the operations of the human brain.

5. Develop the following topic sentences by providing *reasons* or *proof*:

A.

(*a*) The thickness of a grease is no guide to its efficiency as a lubricant.
(*b*) At present, corrosion can be controlled not abolished.
(*c*) Warm air rises.
(*d*) The deterioration of oil is from external causes.
(*e*) Presswork is essential for the production of certain cheap metal articles.
(*f*) A pendulum will not swing perpetually by itself.
(*g*) Animal life is dependent upon the plant world.
(*h*) Toolmakers are key workers.
(*i*) Increased horsepower is not essential for increased mechanical efficiency.
(*j*) Heat is a form of energy.

B.

(*a*) True solids exist in a crystalline state.
(*b*) Voyages to distant stars seem to be beyond the bounds of possibility.
(*c*) Efficient computers were made possible by the development of electronic devices.
(*d*) Air has all the usual properties of matter.
(*e*) The electron microscope has widened the range of scientific studies.
(*f*) Measurement is the foundation of physical knowledge.
(*g*) Human knowledge is cumulative.

(*h*) Space research cannot continue without the expenditure of vast sums of money.
(*i*) Mass-production is a mixed blessing.
(*j*) A local decrease in entropy is always compensated by an increase somewhere else.

6. Develop the following topic sentences in ways that you consider suitable. Specify the method you have used.

A.
(*a*) An engineer is lost without his slide rule.
(*b*) Disc brakes provide greater motoring safety.
(*c*) There are several methods of casting metal.
(*d*) There are different standards of length.
(*e*) The wheel is a special form of lever.
(*f*) Pure iron has very few practical uses.
(*g*) In welding operations, the temperature is critical.
(*h*) A lathe is more versatile than a milling machine.
(*i*) The object of manufacturing is to convert raw material into a finished or semi-finished state.
(*j*) Oil is used in transformers as a cooling medium.
(*k*) Cranes are either of the jib or of the bridge type.

B.
(*a*) A musical note has three characteristics.
(*b*) To produce power we exploit energy originating from the sun.
(*c*) Convection may be regarded as a version of conduction.
(*d*) Work study techniques have been devised to cut production costs.
(*e*) Isotopes have different nuclear masses.
(*f*) Force is a vector quantity.
(*g*) Excessive fatigue saps a worker's concentration.
(*h*) A dam may be built for one of several reasons.
(*i*) Sound travels more rapidly in warm air than in cold.
(*j*) The addition of vectors is achieved by means of simple geometrical construction.

7. The following topic sentences are of a less technical nature. Discuss the possible content of each paragraph, and develop the topic sentences suitably.

(*a*) Communication is a two-way process.
(*b*) The job of a shop steward is not an enviable one.
(*c*) High speed is a factor in the high accident rate.
(*d*) Explosives are not dangerous until you forget they are.
(*e*) The family group is losing its significance.

(*f*) One man overworked can produce more than two men underworked.
(*g*) There is a difference between building and architecture.
(*h*) Oil sludge is a recurring nuisance on our shore line.
(*i*) It pays to advertise.
(*j*) Design and function are closely linked.
(*k*) Television is a mixed blessing.
(*l*) More automation means the problem of more spare time.
(*m*) It is easier to be unreasonable than to be reasonable.
(*n*) Successful productivity depends largely on good personnel management.
(*o*) The romantic days of flying are over.
(*p*) Music is more than 'hi-fi.'
(*q*) People are still very superstitious.
(*r*) Mass production does not ensure mass good taste.
(*s*) You can fool most people most of the time.
(*t*) A man with only his trade is not a complete man.

4. Paragraph Linking

The paragraph is itself part of the larger unit of section or chapter and must, therefore, fit neatly into that unit. It must show some reference to preceding or following paragraphs, perhaps by introducing a series of ideas or by summing up a collection of statements. This means that a paragraph may have not only a topic sentence but also a linking sentence which takes up the thread of previous paragraphs or which states the theme to be developed in the next.

Read the following short exposition on Energy. Note the remarks in the accompanying commentary.

A

The capacity of a body to do work or to overcome resistance to motion is known as its *energy*. Mechanically this energy has two forms—*potential* and *kinetic*.

There is obviously a reference back to previous remarks on 'work.' The paragraph then introduces the term 'energy' in a definition and suggests that the ensuing paragraphs will deal with the two forms.

B

The potential energy of a body depends upon its position. If we raise an object a certain distance above a pre-

The first form of energy is introduced.

scribed level, we can say that it has a potential energy equal to the amount of work it would do in falling to that level. This would be expressed in terms of body weight and height.

This is the first sub-paragraph to paragraph A.

C

Similarly, if a spring is extended or compressed by a certain amount, we can say that it has a form of potential energy equal to the amount of work done in returning to the original position. This is known as *strain* energy and is expressed in terms of the stiffness of the spring and the amount of extension or compression.

'Similarly' is an introductory word leading to a specific type of potential energy. This is defined and explained.

This is a sub-paragraph to paragraph B.

D

Whereas both potential and strain energy depend on the position of the body, kinetic energy depends upon velocity. If a body of certain mass moves with a certain velocity, it possesses kinetic energy because it is able to do a definite amount of work in being brought to rest. Similarly, if a body rotates about an axis, for which it has a certain moment of inertia, at a specified angular velocity, we can calculate the kinetic energy involved in bringing that body to rest. Our result would be expressed in terms of mass and velocity.

'Whereas' introduces a contrast. Kinetic energy is defined and explained.

'Similarly' introduces the idea of rotation, but there is no new paragraph. The distinction between the two types of kinetic energy is not as specific as that between potential and strain.

This is the second sub-paragraph to paragraph A and is parallel to B.

E

If we apply the principle of conservation of energy—that energy can be converted from one form to another but cannot be created or destroyed—we soon realize that the sum of the potential energies and the kinetic energies is constant. Any reduction in potential energy must be accompanied by an increase in kinetic energy. This leads us to the problem of translation of energy.

The two types of energy are drawn together and shown as complementary.

This is a form of summary paragraph, but also shows transition to the next idea, translation of energy, which would be dealt with in further paragraphs.

This short passage could be expressed graphically

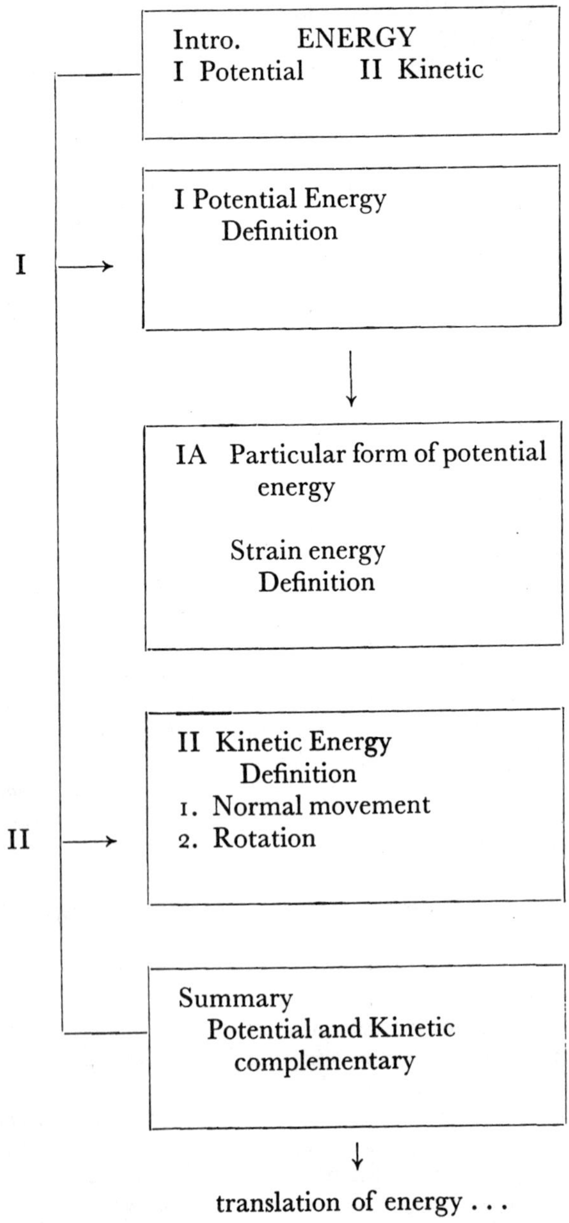

It is sometimes helpful in composing a passage to think graphically and even to scribble a rough blocking of paragraphs to see how they fit together.

N.B. It is important that paragraphs should not be isolated by too complete a break at the paragraph indention. The reader should be carried on by the flow of thought, pausing only for a moment at the rest provided by the indention.

Exercises

Each of the following questions contains a number of topic sentences for you to develop and make into an intelligible and interesting piece of writing. Do not alter the order of the sentences unless you have a good reason for doing so. You may add short introductory, transitional, or concluding paragraphs.

A.

1. (*a*) The first essential of a good weld is correct temperature.
   (*b*) The surfaces to be welded must be perfectly clean.
   (*c*) The welding operation must be controlled so that the action begins at the centre.
2. (*a*) In repetition work a system of inspection is essential.
   (*b*) Various gauges and devices are used to examine parts.
   (*c*) Inspection methods find faults in parts and faults in the production line.
3. (*a*) A lathe can be used to carry out many operations.
   (*b*) The main part of the lathe is the bed.
   (*c*) The headstock is fixed at the left end of the bed.
   (*d*) The saddle rests upon the ways.
   (*e*) The tailstock is positioned at the right end of the bed.
4. (*a*) Of the three types of drill, the flat drill is the simplest.
   (*b*) The straight fluted drill is similar in sectional shape to the flat drill, but has more limited use.
   (*c*) The most commonly used drill is now the twist drill.
5. (*a*) To start a four-stroke engine it is necessary to turn the flywheel.
   (*b*) The mixture is then compressed.
   (*c*) A spark between the points of the sparking plug fires the mixture.
   (*d*) The waste gases are then passed out of the cylinder.
6. (*a*) Fuels may be solid, gaseous, or liquid.
   (*b*) There are both natural and artificial solid fuels.

(*c*) Gaseous fuels have certain advantages.
(*d*) In many engines liquid fuel is essential.

7. (*a*) The element carbon is essential to life.
(*b*) Plant life has given rise to much of our carbon fuel.
(*c*) Carbon is also the basis of more glamorous objects.

8. (*a*) Lubrication is an important factor in bearing efficiency.
(*b*) Gravity feed systems vary.
(*c*) There are several methods of pressure lubrication.
(*d*) Bearings inside gear drives require splash lubrication.

9. (*a*) The work in a lathe can be held in several ways, but a widely used method is between centres.
(*b*) The work may also be held on a mandrel.
(*c*) On such lathes as the capstan and turret, the work is usually held in a chuck.

10. (*a*) A gas turbine consists of two parts.
(*b*) In a turbo-jet aircraft, air is taken in through ducts.
(*c*) The gases are then passed to the turbine blades.

B.

1. (*a*) Taper turning can be carried out by swivelling the top slide rest of the lathe.
(*b*) Another method is by setting over the tailstock.
(*c*) The most satisfactory method is by using a special taper turning attachment.

2. (*a*) The velocity of a body can only be defined in relation to some other body.
(*b*) The term absolute velocity is used when a body is considered as moving relative to the earth.
(*c*) If another moving body is used, the velocity is termed relative.

3. (*a*) Eliminate guesswork in photography by using a reliable light meter.
(*b*) It is important to use the correct developer for the type of film you have exposed.
(*c*) The size of enlargement depends on the quality of your negative.

4. (*a*) A rocket operates according to Newton's Third Law of Motion.
(*b*) It is the same force that . . .
(*c*) The reaction-mass of a rocket must be in small particles.
(*d*) The velocity of a rocket can exceed its own exhaust velocity.

5. (*a*) There are disadvantages in a single-phase circuit.
   (*b*) Two-phase systems have been tried.
   (*c*) The three-phase system is greatly superior to the two-phase.
6. (*a*) A basic task in surveying is the ranging out of straight lines.
   (*b*) Ranging poles are constructed to be seen clearly at a distance.
   (*c*) The poles are subject to certain defects.
7. (*a*) A sliding of one solid body against another in contact is resisted by a force called *friction*.
   (*b*) Friction between dry surfaces follows certain laws.
   (*c*) Friction can be reduced only by the use of lubricants.
8. (*a*) Many modern methods of communication depend upon the photo-electric cell.
   (*b*) A factory may use the principle for burglar and fire alarms.
   (*c*) The photo-electric cell is also useful in laboratories where automatic counting is needed.
   (*d*) The most dramatic use of the cell is in television.
9. (*a*) Measurements of gravitational acceleration are usually denoted by the letter *g*.
   (*b*) The value of *g* varies from place to place.
   (*c*) Acceleration is less on the moon.
   (*d*) Beyond the earth's gravitational pull an astronaut experiences weightlessness.
10. (*a*) The invertor is being increasingly used to replace rotating machinery.
   (*b*) From the simple invertor is being developed a large range of power conversion devices.
   (*c*) There are applications of the devices in aeronautics, marine engineering, and general industry.

5. Planning and Preparation

The most important stage in any written composition is the preparation. The writing of the finished work is only the physical act of using a pen or a typewriter.

Planning an essay or a report does not mean merely jotting down a few disjointed comments against a series of numbers; it involves the arrangement of your material in a logical order, with an obvious beginning, middle, and end. At the time you begin to write you should be perfectly aware of the way in which the writing is to develop; so much so that you should be able to write the last paragraph first, then go back to fill in the preceding ones.

Use the paragraph unit as the basis of your planning. Learn to work out your ideas in terms of paragraphs, then construct a clear framework from those idea-paragraphs so that the task of writing will be straightforward and virtually mechanical. Examine the following method:

1. Collect your material.
2. Select the information you need.
3. Arrange the material in the order of presentation.
4. Construct a framework according to number of paragraphs required.
5. Write down the main topics of each paragraph.
6. Under each topic indicate the development.
7. Construct your topic sentence for each paragraph.
8. Insert transitional paragraphs where necessary.
9. Construct link phrases and sentences if necessary.
10. Make a rough draft.
11. Edit the rough draft and make your final copy.

This may seem to you long-winded, but it is an essential part of any 'apprenticeship' in writing. There is no short cut to writing skill. With practice you should be able to cut down the number of operations by doing some of them in your head.

N.B. Please do not advance the excuse that you might think of something new while you are writing the finished article. If you have any 'brilliant' afterthoughts your concentration in the planning stage must have been poor. Forget them quickly and do not sacrifice the shape and unity of your plan.

Exercises

1. For each of the following topics make an outline plan, indicating the paragraph divisions, with topic sentences and rough notes of methods of development within paragraphs. In A, the number of *main* paragraphs is shown in parentheses at the end of each topic; you may, however, add introductory, transitional, and similar routine paragraphs.

A.

(*a*) Cleaning a car (4).
(*b*) Various types of adhesive and their uses (at least 4).

(*c*) Decarbonizing a motor cycle engine (4).
(*d*) Different methods of house heating (at least 5).
(*e*) The usefulness of plastics (at least 5).
(*f*) A visit to a factory (5).
(*g*) Buying a second-hand car or motor cycle (at least 4).
(*h*) The uses of a library (5).
(*i*) Description of a lathe (5).
(*j*) The uses of X-rays (5).
(*k*) The driving test (4).
(*l*) Building a home workshop (5).
(*m*) Making a simple hot wire ammeter (4).
(*n*) Comparing different methods of visual signalling (5).
(*o*) Problems of lubrication (5).

B.

(*a*) Braking systems.
(*b*) Using a ciné camera.
(*c*) Methods of welding.
(*d*) Mechanical aids in building operations.
(*e*) Road safety.
(*f*) Assembling a 'hi-fi' unit.
(*g*) The uses of the galvanometer.
(*h*) Time-and-motion study.
(*i*) Space probes.
(*j*) Strain-gauges.
(*k*) Motorways.
(*l*) Servomechanisms.
(*m*) Diesel locomotives.
(*n*) Chemicals in the garden.
(*o*) Computers.

2. Using your outline plans, construct and write the *final paragraph only* for each of the subjects.

N.B. The exposition exercises in this book are all exercises in planning and preparation. Before beginning to write any such exercise, you must *think carefully* and *plan on paper*.

6. Style

Exposition—the presentation of facts and information—demands a particular discipline of tone and style. In describing a lathe or an experiment, a speaker has normally no need to raise his voice or gesticulate; he tries to adopt a calm and unaffected tone. Similarly, a writer should attempt to adopt this tone in his use of words and sentences.

A technical writer should not, however, be guilty of considering exposition to be a 'poor relation' of literature. It is impossible to exclude emotion completely from the words he uses—in fact, he would be foolish to try. Engineers and scientists sometimes need to persuade their readers or to use a word for more than its literal meaning, and they should be prepared to go further than the basic principles of clarity, simplicity, and conciseness.

Exposition must flow smoothly and appeal to the reader's sense of rhythm and balance. It is a poor style that delivers long, rambling sentences to bore the reader, or serves up short, choppy sentences to irritate him. A smooth style with variation of sentence length and structure to match the development of the ideas enables the reader to keep up with the argument or the description.

Words can look ugly as well as sound ugly, and, in reading, *both* effects are apparent to the reader. The good writer avoids the clashing of words in a sentence, abhors pompous phrases and sentences, and looks instead for those words and turns of expression which express his meaning and express it in the pleasantest way possible.

Remember the following pointers to a reasonable style:

Try to interest your reader.
Avoid clumsy sentence construction.
Make sure that words do not clash.
Avoid jerky sentences.
Do not let sentences ramble on too long.
Try to write not adequately but *well*.

Personal and impersonal

Language is normally a means of personal expression, communicating what an individual sees or thinks; but scientific enquiry and the need to make writing *objective* has led to a particular style of exposition called 'the impersonal.'

e.g. The liquid was heated in a test tube and a reagent was added

By this means, the writer is careful to omit the personal pronoun 'I,' thereby stressing the objective nature of the experiment.

Beware of accepting this too literally. No amount of 'impersonal' reporting can disguise, or make more scientific, a badly carried-out experiment. Remember that the *facts* and the *information* are important; concentrate, therefore, on expressing them lucidly.

You will no doubt be required, in your technical subjects, to write laboratory and other reports in the impersonal. Remember that there are other styles of technical description, some of which are shown later in this book.

III

DESCRIPTION AND EXPLANATION

These are the two basic forms of exposition. They answer the questions What? How? and Why? What is an oscilloscope? How does a computer work? Why is a rocket the ideal form of transport in outer space?

In answering these questions we have at our disposal the spoken and written word, and the full resources of diagram and illustration.

The important task in describing and explaining is to make sure that your reader obtains as clear an idea as possible of the object or process, and that he fully understands what he reads or sees. To do this you must:

(*a*) know the capabilities and background of your reader
(*b*) be aware of the purpose and limits of the description or explanation
(*c*) select and arrange the material in the best way
(*d*) choose your words correctly
(*e*) be concise in expression
(*f*) be free from ambiguity
(*g*) present the material in an interesting way.

1. Considering the Reader

This is linked with most of the other problems. Knowing your reader's background means that you can draw on *his* experience and use words that will be familiar to him. In this way the chances of success are greatly increased. This does not mean 'talking down' to the reader, or using words of one syllable. If your audience consists of engineers and you are describing a mechanical process you should be able to use technical terms; if the audience is non-specialist, you will have to search for suitable everyday terms in which to phrase your description. This will mean *different* terms, not *simpler* ones.

A knowledge of your reader's capabilities will help you to select the material you are going to use. There is no point in providing him with information that he knows already.

2. Limits of the Description

Many descriptions are vague and ill-defined. Know what you have to describe or explain and keep to the point. If you are required to describe the operation of a lathe, do not wander off into a lengthy digression on milling machines. Introduce the milling action if it is necessary to make a point, but no further.

Your description may be limited by

(*a*) demands of space and time
(*b*) the need for great simplicity (as in a set of instructions)
(*c*) British Standards
(*d*) instructions given by the person asking for the description or explanation.

Be aware of the purpose and decide what your limits are. Take the example of the following variations in the method of describing a typewriter:

(*a*) *as an instructional aid* —written description
parts named and function shown
diagram of parts
limited maintenance advice

(*b*) *as a specification* —dimensions
details of working parts
operation and maintenance
spares available

(*c*) *in a trade journal* —description of main features
emphasis on advantages over other models
test information
comments upon suitability.

(*d*) *in an advertisement* —outline of main features
emphasis of features that have selling appeal

(*e*) *for registering a patent* —formal submission of full details with plans, in accordance with Patent Regulations and British Standards

In the above methods there would be a certain amount of overlapping; but you will see that the purpose and limitations can vary greatly even in describing the same object.

EXERCISE

For each of the following, suggest the various forms a description or an explanation might take. Give your assessment of the type of reader that would be interested.

a centre lathe — a motor scooter — a house
a power drill — a guided missile — a factory fire
centrifugal force — a bonus and incentive scheme
an instruction manual for a washing machine
a 35mm camera — a cement mix — a synthetic fibre
a weed killer — a safety device — a sailing dinghy
an electric shaver — a detergent

3. Arrangement

There are no hard and fast rules for arrangement. The order of presentation depends upon the reader and the limits and demands of the description. These are some of the ways in which particular descriptions or explanations can be arranged:

(*a*) *Description of an object:*

(i) From left to right, following conventional eye movement.

e.g. Describing the general appearance of a machine with detailed references.

(ii) From the general to the detailed, or the reverse.

e.g. Describing the general appearance of a machine before explaining details, fittings, and accessories.

(iii) In the order in which components are assembled.

e.g. Describing machinery for manuals.

(iv) By division into functional parts.

e.g. Describing an internal combution engine in the systems of: ignition; fuel; cooling; lubrication; mechanical.

These arrangements may be enclosed in an overall framework, such as:

I Definition of object

II Description of object (using one of the above methods of arrangement)

III Function of object (scientific principles, etc.)

(*b*) *Description of a process*

(i) Natural sequence of events, in order of time or cause and effect

(ii) Process of manufacture or assembly

(iii) From the simple to the complex. (Leading from the reader's experience to new experiences)

These may be enclosed in an overall framework, such as:

I Definition of process

II Material and equipment required

III Description of process (using one of the suggested systems of arrangement)

EXERCISES

1. Suggest in outline a suitable arrangement for describing each of the following objects. Give reasons for your suggestions.

a lathe	a cantilever	a syphon
an anvil	a bevel gauge	a bunsen burner
a still	a drop-hammer	an outboard motor
a theodolite	a screwdriver	a pressure cooker
a jig-borer	a carburetter	an electric mixer

2. Suggest a suitable arrangement for each of the following processes. Give reasons for your suggestions.

(*a*) Hand-welding two pieces of metal.
(*b*) Building a workshop bench.
(*c*) Making a blueprint.
(*d*) Wiring a coil.
(*e*) Surface grinding.
(*f*) Using a voltmeter.
(*g*) Cold rolling steel.
(*h*) Adjusting the brakes of a motor car.
(*i*) Redecorating the outside of a building.
(*j*) Care and maintenance of a specific piece of equipment.

4. Correct choice of Words

Do not belabour your reader with unnecessary technical terms, especially if they are jargon and your reader is not likely to understand them. The hi-fi enthusiast who talks about 'woofers' and 'tweeters' will communicate successfully with another enthusiast, but his words will be wasted on non-specialists.

Remember that *the correct word is the one which will give the fullest possible meaning in your reader's mind.* If you can use a precise technical term and you know that your reader will understand that term, it is up to you to use it. Above all, do *not* try to impress your reader with your superior knowledge by using long technical words where a simple word will do.

Remember, also, that technical terms have a limited meaning. Sometimes they have been devised and limited by agreement of authorities and commissions for the useful purpose of ensuring that there will be national and even international understanding. This means that there must be no 'near misses.' You must really know your technical vocabulary and use it with accuracy. Be sure that when you use such terms as 'work,' 'energy,' 'conductivity,' 'force,' 'potential,' 'amplification,' etc., you are using them in the agreed way and with the agreed scientific or technical meaning.

Exercises in choosing the correct word are giving later in this Section. Further notes and exercises are given in Section VIII. (Technical Vocabulary) beginning on page (167).

5. Conciseness of Expression

This involves using the right number of words to describe something clearly and without confusion. You achieve this by knowing precisely what you are going to say or write *before* you begin. Afterthoughts and amendments play havoc with your expression.

> *e.g.* Two new benches have been installed in the laboratory, also a new fume cupboard.

This is a typical example of the clumsy afterthought. If you think first you can include the fume cupboard with the benches

> Two new benches and a new fume cupboard have been installed in the laboratory.

Long-windedness of expression can be caused by:

(*a*) *Attempting to seem important*

This often manifests itself (or, to be more concise, 'shows') in the use of long, Latin words. Use Latin or Greek words where they are the agreed labels for chemicals, materials, equipment, formulae, etc., but avoid them if they serve no real purpose.

e.g. Extract a proportion of the requisite allocation of materials.

This is a pompous way of saying

Take such materials as you need.

or

Take what you need of the available material.

(*b*) *Avoiding or forgetting the precise word*

This often causes roundabout expressions such as 'in the course of' instead of 'during', 'due to the fact that' instead of 'because.'

(*c*) *Trying to be impartial or impersonal*

An impersonal style may be required in some forms of exposition such as laboratory reports. It lengthens expression but is sometimes preferable to involving the reader in the process you are describing.

e.g. The next stage is to repeat the experiment using brine.
is preferable to

Now you repeat the experiment using brine.

Do not avoid responsibility by writing impersonally. If you have come to a decision, write:

We should . . .
I recommend . . .

not

It is felt that . . .
It would be advantageous, perhaps . . .

(*d*) *Using outworn phrases*

The descriptive work of many students is cluttered with useless phrases such as 'In so far as . . .,' 'In the case of . . .,' 'As far as . . . is concerned . . .,' '. . . and such like,' which are the result of muddled thinking or laziness.

Examine the following list and determine that these or other forms like them will not litter and disfigure your writing.

With particular reference to
It will be seen that

From some aspects
On the one hand . . .; on the other hand
A certain line of thought
It is probable that
That sort of thing
As I have pointed out
Let us turn to . . .
In certain circles.
From the point of view of . . .
Due to the fact that
Be that as it may
Taking all in all
As a matter of fact
By and large
It is safe to say that
Now then etc.

(*e*) *Clumsy thinking*

This is shown in clumsy expression and in repetition.

e.g. The reason why the machine broke down was because it had not been properly lubricated, a cause which could have been avoided with a little more care and attention.

The sentence shows an uneconomic use of words. It could be written more simply:

The machine broke down because its lubrication had been neglected.

Always *edit* your own sentences and paragraphs to make sure that you have not been guilty of clumsy thinking.

(*f*) *Weak vocabulary*

To cover up vocabulary weaknesses, students often have to use a roundabout expression.

e.g. The finished product was of the quality *laid down in the agreement*.

instead of:

The finished product was of the *stipulated* quality.

Always strive to widen your vocabulary, not with pompous words but with those terms which have been designed to give directness and economy of expression.

Exercises

1. Rewrite the following sentences more clearly and concisely:

A.

(*a*) Due to the fact that the belt drive had snapped it was not possible for the lathe to operate.

(*b*) A type of valve that is in common use today is the one known as the ball valve, where a ball bearing is resting in a seating within the valve and is operated by the force of gravity.

(*c*) There will have to be a certain number of cupboards made available for the staff and with them we will no doubt be compelled to provide desks.

(*d*) So far as lead and copper are concerned they do not dissolve in each other at all and in this direction lead bronzes are just mixes of lead and copper.

(*e*) The reason why the machine which has only just been installed has not been put into operation is because any form of upset is likely to disturb the way in which the present work is now moving.

(*f*) In the case of a blending of moulding powders this is done by the manufacturers and not by the moulders.

(*g*) The lathe is more familiar an object than any other of the machine tools to the general public from an external point of view.

(*h*) One of the great advantages that aluminium has over its fellow metal iron is that for some reason or other it does not seem either to rust or to corrode.

(*i*) As a result of the fact that we have grown so used to living within an atmosphere the perception of air pressure is something that we only realize when the pressure affects us in some way, perhaps by the uncomfortable effects at high altitudes.

(*j*) The modified slotting machine, which has had several additions and alterations made to it, will be tested in a few days time and the results will indicate whether or not the modifications have been worthwhile.

B.

(*a*) In so far as molecules are concerned, all those which go together to make a pure substance are exactly the same and the difference between one substance and another is to all intents and purposes that between the types of molecule.

(*b*) The job was set up, with a driving dog clamped on the workpiece using a piece of sheet copper or brass under it to avoid damage to the moved surface, between centres.

(*c*) As a consequence of the fact that there is uneven heating of some parts of the surface of the earth there is a movement of the air caused mainly by hot air rising and being replaced by air which has come from a cooler region.

(*d*) One way of describing an electric current would be to say that it is a movement of electric charges along a path which can be defined. The path just mentioned is, in the case of an electric current passing through a copper wire, defined by consideration of the geometry of the copper conductor.

(*e*) In this modern day and age, much of the wonder of flights by rockets and missiles depends largely on the scientific process known as *telemetry*, which might be defined as the technique of measuring, transmitting, and receiving information obtained at a long distance by electronic methods, and also the technique of integrating and recording that information.

(*f*) The bubble which indicates, in a spirit level, the correct setting for a level surface or plane is gradually being replaced by another, more reliable method which involves the use of a system of swinging prisms. Such a method of obtaining the correct level is simpler to carry out and the level is self-aligning, another great advantage.

(*g*) A cathode ray tube comprises a glass tube of a certain design within which a narrow beam of electrons can be controlled and directed so that they land on a screen which has been coated with a certain substance which becomes fluorescent when it is struck by this beam of electrons. The result of this contact is that a spot of light is thus formed and this spot can be controlled and moved all over the screen in any way by deflecting the beam of electrons.

2. Rewrite the following sentences replacing each of the phrases in italics by a single word that conveys the same meaning.

A.

(*a*) He asked for an *idea of how much the work would cost.*
(*b*) The lorry *went backwards* into the wall.
(*c*) The light switch was *fitted so that it did not stick out from the wall.*
(*d*) You should *do* the process *in exactly the opposite way.*
(*e*) This is a thrust *operating in a crosswise direction.*
(*f*) The garage mechanic carefully greased the suspension members to *get rid of* the rattles.

(*g*) The noise of the compressor was *going on all the time.*
(*h*) He *admitted light to* the film for 30 seconds.
(*i*) The marking-out table is of cast-iron and is supported *in a parallel plane to the ground* on rigid legs.
(*j*) A dressing used on leather drive belts will keep them *soft and easily bent.*
(*k*) The setting up of a round bar in a four-jaw chuck is apparently *regarded as being a hard piece of work.*
(*l*) The *rod with the weighted end* swung quietly and regularly.
(*m*) They *made partial changes in* the engine design to improve the performance.
(*n*) In some petrol engines the mixture of gas and air is sparked by a *kind of revolving induction coil.*
(*o*) It was impossible to obtain spare parts for that type of machine as it was *now completely out of use.*

B.

(*a*) The substances were *of plant or animal origin.*
(*b*) Rocks *formed from mud or sand banks* hold hidden wealth for the scientists who explore them.
(*c*) For heat insulation of the metal in the rocket nozzle they used linings *made from a pottery material.*
(*d*) Substances *which have the power of emitting invisible rays that penetrate opaque matter* can be traced by means of special 'counters.'
(*e*) After long periods at high temperature, turbine blades suffer from *continuous deformation and elongation of the metal.*
(*f*) When preparing and servicing the rocket for launching they use a *large crane-like structure.*
(*g*) He ground the glass until it was *curved like the outside of a sphere.*
(*h*) The queer mountain shapes were the result of constant *action of wind and rain.*
(*i*) The re-wiring of the house was delayed by the shortage of *pipe for containing the electric wires.*
(*j*) For some distance the road and the railway track ran parallel; then they *began to spread apart.*
(*k*) They chose that particular type of stone because it was *able to resist the passage of water.*
(*l*) The armature winding involved the use of a conductor *made up from a number of thin strips.*
(*m*) The use of a fixer is a part of the developing process *which is essential to its complete success.*
(*n*) The effect of the drug upon the patient was *only for a very brief space of time.*

(*o*) The foreman discovered quickly that the new employee was *not able or qualified to do his work.*

(*p*) The problem *which is at the basis of this matter* is that there is insufficent floor space.

(*q*) When the shop was painted the manager began to see how the colours he had thought would clash *lost their distinctiveness by fitting in* with the other decorations.

(*r*) The *man who looked after trade union business in our workshop* was *serving as the means of* obtaining better working conditions.

(*s*) Some modern scientific achievements have been made in spite of difficulties that seem *to be impossible to overcome.*

(*t*) They calculated the conditions *which were most favourable* for reproduction of that species.

6. Freedom from Ambiguity

It is very easy to make statements that are capable of more than one interpretation. Ambiguity is, after all, a fundamental part of our humour and a strong element in the structure of literature, particularly poetry; but it is a grave danger in technical writing where the aim is not to broaden the meaning and significance of a word and statement but to limit the use of the word so that it is completely clear of any possible misunderstanding or misinterpretation.

Ambiguity often provides people—particularly English lecturers—with a fund of unconscious humour.

e.g. There will be trouble in this workshop—the men take things too easily.

Are the men lazy or downright dishonest?

Ambiguity has a number of causes, including:

(*a*) misplacing of phrases

e.g. He made a cigarette lighter for his friend on the lathe.

(*b*) unfortunate dual significance of a word

e.g. The work was done at his convenience.

(*c*) dual significance of a particular phrase

e.g. That was the last job he wanted to do.

(N.B. This is clearer in *spoken* English)

(*d*) misplacing of the adverb in a sentence

e.g. The machine needed oiling *badly.*

(*e*) faulty punctuation

e.g. Above the first floor offices were occupied.

(What happens when you put a comma after 'above?')

(*f*) ambiguous pronouns

e.g. The line-manager told the foreman that he and not his assistant would take responsibility for distribution of work and that in future all complaints would be referred to him.

N.B. Always edit your work and do not let ambiguities slip through.

EXERCISE

Explain what constitutes the ambiguity in each of the following sentences and try to rewrite the sentence to convey what was apparently meant.

A.

(*a*) The box contains 14 articles weighing three kilogrammes.
(*b*) There is heavy rainfall in Britain generally at this time of year.
(*c*) Nobody has been in the factory for a week.
(*d*) Approximately half of the staff work in this room.
(*e*) All apprentices must park motor cycles in car park D.
(*f*) He used a voltmeter to check the circuit only to find that it was faulty.
(*g*) If you write to us later this year we will arrange to show you round our works as quickly as possible.
(*h*) He was known to move in eccentric circles.
(*i*) The foreman told Bill that he would be moving to the new factory.
(*j*) He took a photograph from the window which was over-exposed.
(*k*) It is difficult to operate a machine with a damaged arm.
(*l*) He almost operated the machine single-handed.
(*m*) He watched the process carefully taking down notes.
(*n*) They allocated the machine to the workman with the new safety attachments.
(*o*) The members of the next discussion group are pinned on the noticeboard by the main entrance.

B.

(*a*) Throughout the business planning has been decidedly haphazard.
(*b*) The assembled members condemned the stoppage by a large majority.

(*c*) Sports cars only appeal to some motorists with noisy exhausts.
(*d*) A bonus and incentive scheme was introduced into the workshop which was not popular.
(*e*) If we mix limewater with carbon dioxide it will turn milky.
(*f*) The new machine was contained in a recent copy of the house journal.
(*g*) The ends of the motorways will remain either where they are now, in the countryside, or local authorities must take giant steps to clear a way for them through property.
(*h*) He asked him when he had made the arrangements for servicing the machines.
(*i*) He asked one of the office staff to put on the electric fan because it was hot.
(*j*) This is the camera with the attachment on the accessory shoe that enables the photographer to compensate for parallax. Should it prove ineffective during the first stages of use it should be taken to the nearest dealer who will deal with it.

7. Metaphor

Metaphor is the term given to the use of a word in other than its normal significance. It is a form of transfer to create an effect, as in the simple example of the verb *plough* which may be used metaphorically:

The ship *ploughed* through the waves.

The metaphor here depends upon the resemblance between the action of the ship in a seaway and the action of a plough upon the soil. Note how this important factor of resemblance occurs in the following metaphorical statements:

a *biting* wind a *flood* of tears
Everything *hinged* on his action

Metaphor is basically of two kinds—living and dead. Living metaphors are those we use knowingly and which we appreciate for their presentation of resemblances. Dead metaphors are those to which we have ceased to react and which occasion surprise only when some one points out that they have a metaphorical origin.

e.g. *laws* of science *saddle* of a lathe
choke control *feeding* a cutting tool

Metaphors of a literary kind are usually out of place in technical writing, and there are occasions when their use can be positively dangerous. Even some live metaphors have become doddery in their old age, and no writing is given vitality by the use of such terms as 'leaving no stone unturned,' 'The road to success,' etc.

Dead metaphors, such as *law* of nature, the *energy* in an atomic reactor, are also dangerous if the reader accepts them in more than their scientific sense and tries to bring the metaphor back to life.

It is obvious that any use of metaphor in technical exposition must be careful and necessary and should help to convey the meaning without obscuring the facts. A good metaphor enhances the style and the reader's interest, and provides a freshness of approach; but it should also bear a mark of familiarity.

'*Sift* the information' can be expressed in non-metaphorical terms; but can this be done as effectively? 'closely examine,' 'analyse carefully,' 'examine in detail'—these expressions do not go as directly to the point nor are they as effective in *conveying* the information to the reader.

8. Analogy

Analogy can be an important technique in description and explanation. We use the term to indicate the explanation of a process by stressing its similarity to another process more familiar to the reader. For example, the Earth can be regarded as being in a kind of steep-sided gravitational pit out of which a rocket missile must climb before it can start over the almost level gravitational plains leading to other planets. The analogy here is between the problem of leaving the earth's gravitational pull and that of climbing out of a deep pit.

The great danger of analogy is that the reader may regard the resemblance shown as being an actual parallel rather than a means of moving from the known to the unknown.

Use analogy sparingly and in accordance with the needs of your reader. The non-specialist reader will obviously benefit more from the use of analogy than will the specialist.

e.g. To say that 'The new electron microscopes are capable of magnification by 1,500,000' may impress some people, but to many it is just a large figure.

Analogy can vivify the information for the layman:

> '. . . magnification such that if a penny were so magnified it would cover almost the whole of London.'

EXERCISE

Suggest a suitable analogy to describe the following. Where possible expand your suggestion into a short paragraph.

sound waves	a fulcrum
molecules	the nervous system
rocket propulsion	glass-blowing
mixing cement	factory workers
a camera	memory
a catalyst	magnetic attraction
polarizing light	radar
centrifugal force	

9. Illustration

This is vitally important in technical description and explanation. At the completely practical level, an engineering drawing or a blueprint may be all that is required—words would be superfluous. Sometimes the illustration may need explaining; in a lengthy report giving conclusions and recommendations illustrative material may be assembled at the end, in the appendices.

All illustrations must be up to the technical standards required, with the correct lettering and references, and they should contain only the information relevant to your purpose.

10. Maintaining Interest

Being concise and factual does not mean being terse and abrupt in style. Professor Kapp[1] provides a useful rule: 'Add to the words needed to express the information as many more as may be needed to *convey* it.' A description should flow smoothly and the linking words should show the development of the writer's thoughts.

The cramming of facts and descriptive material into a small compass causes mental indigestion, and is in no way the same thing as being *concise*.

[1] *The Presentation of Technical Information.* R.O. Kapp (1948).

Interest is mainly infectious. If you can show your reader how interested *you* are, it will not take him long to join in.

Exercises in Description and Explanation

1. Describe the following equipment with the aid of illustrations:

A.

(*a*) dynamo
(*b*) hand power drill
(*c*) sparking plug
(*d*) vice
(*e*) tubeless tyre
(*f*) callipers
(*g*) hacksaw
(*h*) bunsen burner
(*i*) electric kettle
(*j*) ball valve

B.

(*a*) surfacing machine
(*b*) pillar drill
(*c*) reduction gear
(*d*) pneumatic drill
(*e*) hot wire ammeter
(*f*) glider
(*g*) hydraulic press
(*h*) microphone
(*i*) pile-driver
(*j*) simple microscope

2. Draw diagrams of the following. Letter or number the parts and compose a brief description using those references.

A.

(*a*) syphon
(*b*) tyre pressure gauge
(*c*) grease gun
(*d*) electric light bulb
(*e*) vacuum flask
(*f*) stapler
(*g*) syringe
(*h*) thermometer
(*i*) pantograph
(*j*) lawn-mower

B.

(*a*) cathode ray tube
(*b*) capstan lathe
(*c*) globe valve
(*d*) tumbler switch
(*e*) carburetter
(*f*) ciné camera
(*g*) micrometer
(*h*) cement mixer
(*i*) hydrometer
(*j*) amplifier

3. Describe the following *without* the aid of any drawing or diagram.

A.

(*a*) ring spanner
(*b*) pair of pliers
(*c*) spirit level
(*d*) tenon saw
(*e*) contraction rule

B.

(*a*) blowlamp
(*b*) boat's compass
(*c*) slide rule
(*d*) simple strain gauge
(*e*) guillotine

4. Draw diagrams of the following items, but do not use letter or number references. Then write a suitable description.

A.

(*a*) box camera
(*b*) arc lamp
(*c*) thermostat
(*d*) electric motor
(*e*) dumpy level
(*f*) rheostat
(*g*) alarm clock
(*h*) crane
(*i*) telephone receiver
(*j*) shear-legs

B.

(*a*) flash gun
(*b*) power hammer
(*c*) rifle
(*d*) weaving loom
(*e*) inclinometer
(*f*) vacuum cleaner
(*g*) escalator
(*h*) bull-dozer
(*i*) aqua-lung
(*j*) thermo-electric pyrometer

5. The following exercises are in the description or explanation of *processes*. The initial exercises in category A are divided into specialist sections.

A. (*Metalwork*)

(*a*) Filing a piece of metal.
(*b*) Sharpening a cutting tool.
(*c*) Brazing.
(*d*) Setting up a tool in a lathe.
(*e*) Chilling iron.
(*f*) Moulding a flanged pipe.
(*g*) Beating a piece of copper into a concave shape.
(*h*) Scraping a piece of metal after machining.
(*i*) Centring and drilling the ends of a piece of work.
(*j*) Annealing.
(*k*) Buffing a piece of metal.
(*l*) A method of case-hardening small steel parts.
(*m*) Hand forging.
(*n*) System of feeding coolant to a cutting tool.
(*o*) Boring a tapered hole by means of a lathe.

(*Electrical*)

(*a*) Operation of a simple electric motor.
(*b*) Thermostatic control.

(*c*) Constructing a simple transistor circuit.
(*d*) Tracing a fault in a circuit.
(*e*) Operation of a magnetic amplifier.
(*f*) Wiring a house on the ring main system.
(*g*) Simple electro-plating.
(*h*) Making a tape-recording.
(*i*) Measuring levels of noise.
(*j*) Using a moving-coil meter.

(*Building and Civil Engineering*)

(*a*) Making reinforced concrete piers.
(*b*) Using a prismatic compass.
(*c*) Laying bricks in Flemish bond.
(*d*) Laying a damp course.
(*e*) Constructing a retaining wall.
(*f*) Testing a beam.
(*g*) Measuring with a surveying chain.
(*h*) Laying a cement or asphalt path.
(*i*) Constructing a simple brick archway.
(*j*) Testing a drainage system.

(*Motor Vehicle*)

(*a*) The operation of a two-stroke engine.
(*b*) Setting a carburetter.
(*c*) Adjusting brakes.
(*d*) Balancing the wheels.
(*e*) Adjusting the track.
(*f*) Replacing a defective water pump.
(*g*) Adjusting the timing mechanism.
(*h*) Changing the points.
(*i*) Tracing an electrical fault.
(*j*) Decarbonizing an engine.

(*Miscellaneous processes*)

(*a*) The action of a force pump.
(*b*) Experiment to find the density of water.
(*c*) Measuring an internal diameter.
(*d*) Drawing down a rectangular bar.
(*e*) Using a testing plate.
(*f*) Strength testing of rubber or similar compounds.
(*g*) The principle of the hydraulic press.
(*h*) A photo-copying process.
(*i*) Drawing an isometric projection.
(*j*) Tempering steel.

B. The following processes are more complicated and may require a series of diagrams linked by carefully composed commentary:

(*a*) Stroboscopic inspection.
(*b*) Operation of a resistance furnace.
(*c*) Operation of disc brakes.
(*d*) Electric arc welding.
(*e*) Operation of a smoke tunnel.
(*f*) Frequency modulation.
(*g*) Reducing iron ore to pig iron.
(*h*) Sand-casting of metal.
(*i*) Generating electricity in a power station.
(*j*) Operation of a turbo-jet engine.
(*k*) Construction of a suspension bridge.
(*l*) Operation of a simple assembly line.
(*m*) The Bessemer process.
(*n*) Re-boring and re-sleeving an internal combustion engine.
(*o*) Seismographic oil prospecting.
(*p*) Inspection of aircraft sections in a supersonic tunnel.
(*q*) Operation of a hovercraft.
(*r*) Construction of a tarmacadam road.
(*s*) Making a small boat.
(*t*) Drop forging.

6. *Explain* carefully and objectively the following phenomena:

A.

(*a*) lightning
(*b*) a mirage
(*c*) smog
(*d*) an echo
(*e*) an explosion
(*f*) heating by radiation
(*g*) sunstroke
(*h*) parallax
(*i*) cumulus clouds
(*j*) eclipse of the moon

B.

(*a*) refraction
(*b*) volcanic formation of a round crater
(*c*) a hurricane
(*d*) radioactivity
(*e*) the sound barrier
(*f*) weightlessness in outer space
(*g*) aerodynamic instability
(*h*) capillary action
(*i*) earthquakes
(*j*) Coriolis force

IV

SPECIAL TYPES OF EXPOSITION

GIVING AND WRITING OF INSTRUCTIONS

1. Oral Instructions

These are more difficult to deliver than many students realize. So many instructors will credit their listeners with the power of mind-reading.

Always consider the following points:

(*a*) What is the background of your audience?
Will they understand the words you use, and have they the ability to understand *all* that you are saying?

(*b*) Are you giving the instructions at a suitable time?

(*c*) Are the instructions so worded that they are not only easy to understand but easy to remember?

(*d*) Have you allowed sufficient opportunity for the listener to question and confirm?

(*e*) Are you bothering to confirm that your instructions have been 'received and understood'?

Effective instructions may include demonstration and illustration—blackboard or pencil and paper—and the technique of careful repetition to ensure complete understanding and absorption of the information.

EXERCISES *Group Practice*

1. The student group is provided with paper and pencils. The lecturer, or one of the students, sits at a desk in front of the group with a piece of drawing paper concealed from the group's sight.

Another student, acting as 'instructor,' stands by the desk where he can see the drawing paper clearly.

The lecturer then draws a pattern or diagram, pausing between the stages of construction to allow the 'instructor' to tell the group what they must do to get the same result. This must be done without gesture or use of blackboard, and no reference back or questioning is permitted. The group must make their drawings precisely as instructed.

When the diagram has been completed the lecturer compares it with those produced by the group, and any differences are analysed in the light of instruction failure.

This exercise may be varied by allowing questions and recapitulation.

2. Practise within the group the oral delivery of the following instructions. Where possible, listeners should carry out the process on the spot. With longer processes, the work can be done later after preparation has been made from the oral instructions in the form of rough notes.

(*a*) Tying a shoe lace or necktie.
(*b*) Fitting a chuck to the headstock of a lathe.
(*c*) Drilling and countersinking a hole (specify dimensions).
(*d*) Developing a roll of film in a tank. (Use a dummy roll in the light.)
(*e*) Changing the slow-running jet in a carburetter.
(*f*) Wiring a two-way light switch.
(*g*) Building a small crystal set from components.
(*h*) Cutting an external thread on a bar, using a die, and specifying dimensions.
(*i*) Making a mortice and tenon joint.
(*j*) Re-wiring the armature of an electric motor.
(*k*) Maintaining and servicing a motor lawn-mower.
(*l*) Operating a calculating machine.
(*m*) Testing an electric circuit for shorting.
(*n*) Measuring the resistance of a coil.
(*o*) Taking measurements with a theodolite.

2. Written Instructions

Much the same rules apply; but this is 'long-distance psychology.' It is much more difficult to assess the reader and so create the right conditions for understanding. You should, however, be able to estimate roughly the type of reader for a particular set of instructions. For example, an instruction manual for a motor vehicle usually takes two forms:

(*a*) a detailed explanation of assembly, repair, and maintenance.
(*b*) a general outline of the main parts and their day-to-day care and maintenance.

Obviously (*a*) is designed for garage mechanics and (*b*) for the owner-driver. The approach, vocabulary, and content will be different, even though *some* owner-drivers may have sound mechanical knowledge.

In framing written instructions you must make sure that no snags appear to the reader. You must answer possible questions first, because there is little chance of the reader being able to ask them. Put yourself in his place and examine your own work ruthlessly. Do not expect him to be a thought-reader.

Arrangement All the steps in the process must be shown clearly and usually in the order in which they are to be done. If there are certain parallel operations this fact should be indicated *clearly* at the beginning of the set of instructions, and, if possible, an outline summary of the process provided.

Paragraphing Usually instructions are written in short paragraphs to enable the steps in the process to be seen clearly. Long paragraphs tend to create confusion and to wreck concentration. Enumerated paragraphs are helpful:

E.g. 1. Open the stopcock. . . .
2. Remove the. . . .

Style of presentation

The following methods are used:

(*a*) *Impersonal style*—The casting is removed from the mould by the operator, who then places it. . . .
(*b*) *Indirect style*—The operator removes the casting from the mould. He then places it in. . . .
(*c*) *Direct style*—You remove the casting from the mould. You then place it in. . . .
(*d*) *Imperative style*—Remove the casting from the mould. Then place it in. . . .

The relative merits of personal and impersonal style are dealt with in Section II, page 58. In the following exercises use your own discretion as to the style you use, then discuss the different approaches within the group.

Diagrams

These are essential parts of the process of instruction and they should be used carefully and marked clearly. Be selective in your use of diagrams (plan, elevation, etc.) to suit the particular material.

EXERCISES

In the following exercises plan your work carefully, making rough drafts if necessary. Give prime consideration to the potential audience—some indication of a possible audience will be given in certain exercises. Vary your framework and method of presentation as you consider suitable.

1. The following is a set of instructions for the method of assessing a ship's speed in the open sea. Convert the passage into a set of *direct* instructions, with each stage clearly marked and numbered. Use diagrams if necessary.

The Dutchman's Log

This is an accurate method of finding the speed of the ship at a given moment, when out of sight of land. It can be used advantageously during power trials in the open sea.

Before the trial select two observation stations on the upper deck, one forward and one aft. The horizontal distance between the stations, which must be accurately measured from the plan of the ship, should be as great as possible. At each station a fixed open sight should be set up and trained accurately on the beam. If possible, each station should be connected by buzzer to a control position. Numerous floats of suitable shape and size should be prepared, and, if possible, a catapult should be constructed for throwing the floats from the forecastle. The forward observation station must be sufficiently far aft to allow the floats to reach the water before they pass the line of sight. If arrangements can be made for other ships to drop the floats from the position ahead, so that the floats will pass the observing ship at about three-quarters of a cable on the beam, the observation stations can be nearly the whole length of the ship apart.

When all is ready a float is dropped from the ship ahead or catapulted from right forward as far as possible on the bow. At the moment the float passes the line of sight, the observer at each station presses his buzzer key and both times are noted to 1/5th second at the control position. About five floats, dropped at short intervals, are required for each set of observations; and the mean of the speeds obtained from the five observations should be taken to be the speed of the ship during the period covered by the observations. Great accuracy in observing and taking times is necessary to obtain results in a fast ship; for example, if the observers are 167 yards apart and the speed of the ship is 30 knots an error of 1/5th second in the elapsed time will produce an error of 0.6 knot in the speed obtained from one observation.

Admiralty Navigation Manual, Vol. I

2. Set out below is a description of cross-filing. Convert the passage into a set of direct instructions, using the imperative and arranging the stages in numbered paragraphs. Make your instructions clear by using diagrams.

The work should be held firmly in the vise with the minimum amount projecting, and with the surface to be filed truly horizontal. The file handle is grasped in the right hand . . ., the end of the file handle pressing against the palm of the hand in line with the wrist-joint. The left hand should be used to apply pressure at the end of the blade. . . . For lighter filing a rather lighter grip is used with the right hand and the end pressure is applied with finger and thumb. . . . A position should be taken up on the left side of the vise and the feet firmly planted, slightly apart. A stroke should be made by a slight movement of the right arm from the shoulder, and by a sway of the body towards the work, each of these movements being about equal. As the file moves over the work it does not move parallel to its length, but in an oblique direction from left to right. The file must remain horizontal throughout the stroke, which should be long, slow and steady, with pressure applied only on the forward motion. On the return stroke, although the pressure is relieved from it, the file remains in contact with the work.

Success in filing flat is dependent on keeping the file horizontal throughout its stroke, and this is controlled by the distribution of pressure as between the two hands. The fault with beginners is to apply too much downward pressure with the right hand at the beginning, and with the left hand at the end of the stroke, causing the file to rock, and produce a round instead of a flat surface. Careful practice in gradually shifting the pressure from the left to the right hand will ultimately bring success in the production of a flat surface. But even with the best regulated filing there is always a tendency to rock the file slightly, and the curved surface of a tapering file tends to make a hollow surface, thus counteracting any slight rocking.

To test the surface of work during filing a straightedge should be placed on it occasionally, and the line of contact viewed for 'daylight.' When any considerable amount of metal has to be removed, the bulk of it should be removed by a rough or bastard-cut file and the surface progressively brought to a finish by second-cut and smooth files.

Workshop Technology, Part I, by W. A. J. Chapman

N.B. Do not be afraid to alter the wording or the order of the original. Your set of instructions should be briefer than the description and should be in chronological sequence.

3. Most people like to talk about their hobbies, but how well can they pass on information and instructions about them. Carefully prepare and write a simple set of instructions for one of the basic techniques involved in your hobby. The following list contains some suggestions:

(*a*) *Camping:* pitching a tent; open-air cooking; cleaning camp.
(*b*) *Underwater Swimming:* selecting equipment; safety precautions; making the first dive.
(*c*) *Sailing:* rigging and launching a sailing dinghy; getting under way; elements of sailing.
(*d*) *Fishing:* choosing the equipment; fly-fishing; sea-fishing.
(*e*) *Athletics:* training routine; competitive running.
(*f*) *Photography:* using a camera; developing a film; using a flash gun; technique of lighting.

Instructions on choice of equipment, safety factors, and elementary techniques can be prepared for such subjects as climbing, cycling, model-making, 'karting,' and stamp-collecting.

4. Give clear concise instructions for the following simple operations. Use diagrams if you wish. Select the style of presentation.

(*a*) to (*e*) are intended for non-technical readers.
(*f*) to (*j*) are intended for readers with a basic engineering training.

(*a*) Replacing a blown fuse.
(*b*) Charging a car battery with a trickle-charger.
(*c*) Repairing a puncture in a motor-car tyre (tube or tubeless).
(*d*) Various methods of removing a nut and bolt which have set hard with corrosion.
(*e*) Developing and fixing a photographic film.
(*f*) Replacing a damaged cylinder gasket (specify engine type first).

(*g*) Splice-welding two iron bars.
(*h*) Measuring a piece of work with a micrometer.
(*i*) Setting up a piece of work in a lathe.
(*j*) Decarbonizing a motor-cycle engine.

5. Give clear, concise instructions for the following. Audience is specified at the end of each topic.

(*a*) Fitting and operating an outboard motor (*general public*).
(*b*) Finding the surface area of a cone (*technical students*).
(*c*) Hot-riveting two metal plates together using countersunk rivets (*technical students*).
(*d*) Constructing and using a simple pin-hole camera (*schoolboys*).
(*e*) Enlarging, developing and printing a 9 × 14 cm print from a 35 mm negative (*amateur photographers*).
(*f*) Adjusting the tappets on a four-cylinder o.h.v. engine (*owner-drivers*).
(*g*) Threading the film in a 16 mm projector (*non-technical group*).
(*h*) Use of a vernier protractor (*technical students*).
(*i*) Examining machinery for wear with a stroboscope (*technical students*).
(*j*) Testing the valves in a radio receiver (*technical students;* specify type of equipment).

6. Design a set of short operating instructions suitable for accompanying or being attached to each of the following machines. Remember that you must use the absolute minimum of words.

pillar drill	bending machine	power press
milling machine	simple wind tunnel	light meter
prismatic compass	spray gun	tape recorder
oscillograph		

7. Design care and maintenance cards for each of the following machines. Show only basic C and M, with intervals.

motor cycle	drilling machine	typewriter
sailing dinghy	rifle	tape recorder
generator	rotary cultivator	lathe motor
cement mixer		

LETTER-WRITING

Much communication in business is still by letter and, although the typing and sending out of letters is largely an office task, an increasing number of engineers are filling the top executive positions and have to face the problem of dictating letters.

At this stage it will be as well for you to remember the following important points:

(*a*) A letter is written in *English* and not in that hybrid language sometimes called 'commercial English.'
(*b*) The important person is the one who is going to *read* the letter.
(*c*) A letter has a 'shape.' It must look neat and presentable upon a typed page.
(*d*) Every letter that goes out from an organization is a 'letter of goodwill.'

Layout of a Business Letter

The following sample typed letter gives an idea of how the material should be displayed on the page.

MARTIN AND BROWNE
Engineering Consultants

Tel. MAYfair ......

Holt House,
Swift Street,
London, W.1.

Our Ref: PV/SM 15th May, 197-

The Manager,
Gladwell Engineering Co. Ltd.,
Bell Street,
Torbury.

Dear Sir,

Castings Type P.73 H.

On the 10th May you supplied us with 200 malleable iron castings, Type P.73 H., needed for our contract with the Hartmund Development Corporation. Unfortunately, 30 of those castings were imperfect and impossible to use in the construction of the pumping equipment.

Will you please supply replacements within the next seven days, to prevent undue delay in the contract work?

We are returning the defective castings by rail.

Yours faithfully,

R. M. Wilson
Assistant Engineer

(*a*) *Letter heading.*	Usually printed.
(*b*) *Date*	Always given in full, in the form shown.
(*c*) *Reference*	For filing and reference purposes. Usually initials of person dictating followed by those of shorthand typist or secretary.
(*d*) *Inside Address*	Address of person to whom letter is being sent. This is the position in a business letter, but in some formal letters the inside address is at the bottom left of the letter.
(*e*) *Salutation*	Formal way of addressing the reader. Varies according to relationship, and is linked in style with the *complimentary close.*

Type of letter	*Salutation*	*Complimentary Close*
Official and	Dear Sir, Dear Sirs,	Yours faithfully
formal	(Dear) Madam, Mesdames	Yours truly
Semi-personal	Dear Mr. . . .	Yours sincerely
	Dear Mrs. . , etc.	Yours truly

(*f*) *Subject Heading*	Helpful to the reader.
(*g*) *Body of the Letter*	Must be written in clear English. Paragraphing is usually short and with a single topic. Punctuation should be used to ensure an exact, unambiguous reading.

Style

The writer should aim at a positive *tone*—polite and definite—avoiding pompous language and technical jargon, particularly when the letter goes from the firm to a member of the public. Words such as please, appreciate, grateful, are words with a definite, pleasant tone.

BE CLEAR, COURTEOUS, and SINCERE

Jargon

Avoid commercial jargon of the following kind:

I am in receipt of your letter (*Thank you for your letter*)
It is understood to be necessary that (*We must*)
In re your favour of (*With reference to your letter*)

Avoid:

1. *Clichés such as* Believe me to be
Thanking you in anticipation
I remain
Hoping to be of assistance
In due course

2. *inst., ult. and prox.* Their use is still common, but not sound practice. There is always the chance of confusion. Better to put the date in full.
inst. (instant) this month
ult. (ultimo) last month
prox. (proximo) next month

3. *The crude imperative* A bad tone is created if you phrase your requests in the imperative.

E.g. We can supply a variety of standards. Fill in the enclosed form and....

'Please' before 'fill' turns the curt instruction into a polite request.

N.B. THINK MORE OF THE READER THAN OF YOUR OWN IMPORTANCE.

EXERCISES

1. Convert the following examples of jargon into good English.
(*a*) It is regretted that you are dissatisfied with the work done by us.

(*b*) It is incumbent upon us to point out that further trouble may develop.

(*c*) It has been brought to my notice that the upward trend in oil prices has been coincident with an upward rise of prices in other commodities.

(*d*) Consequent upon your order of recent date, it will be our earnest endeavour to fulfil your order with the utmost despatch.

(*e*) Dear Sir,

Your delivery of the 15th March has not yet been received by us, and it is felt to be necessary that some definite action be taken. The appropriate office of British Railways will be contacted, and every endeavour made to trace the missing machinery.

In the event of the consignment being recovered it will be our first duty to communicate with you to that effect. We shall then arrange to collect at your convenience.

2. Draft and set out well-written letters in accordance with each of the following brief instructions. Be particularly careful with paragraphing.

(*a*) Write to your local printer asking him to print a series of safety-first posters for you. Specify layout, size, number of copies and other relevant information.

(*b*) You have ordered, for your firm, a supply of tractor tyres. The letter was sent to the suppliers two weeks ago but you have received no reply. Write a suitable (polite) letter to the tyre firm.

(*c*) Reply suitably and tactfully to this letter:

Dear Sir,

I can't sleep a wink at night. The noise from the machinery in your factory is dreadful. Do you have to work all through the night? Why can't you fit a silencer?

Yours faithfully,

P. B. Jones

(*d*) Write a letter of confirmation of appointment to the man you have just interviewed for the post of Assistant Storekeeper. You should tell him when and where to report, and what documents to bring, and confirm that his holiday arrangements will be respected.

(*e*) You have been asked to give a short talk to a local society. They have suggested something on micro-wave theory, but this is not the subject on which you would be happy to lecture. Draft your reply.

(*f*) An old employee has now reached the age when he should retire. He is, in fact, over the age, and a stubborn old fellow. Write a persuasive letter pointing out to him why he should retire.

(*g*) As manager you have been asked for a confidential report on one of your employees. The letter has come from a firm which is considering his application for a new job. This is the first you have heard of this and you are not very pleased with the man's work. Draft your reply.

(*h*) As shop steward of the machine shop write to the manager asking for special permanent facilities for union meetings.

(*i*) As sales manager of a firm making automatic venders, draft a letter to a large engineering works outlining a special offer of loan of equipment on approval.

(*j*) Apologize to a customer for delay in delivery of certain items of equipment. A breakdown in one of your shops had hampered production.

(*k*) Draft a letter of enquiry to a firm of package manufacturers. You wish them to design a type of case for shipping new, delicate electronics equipment to tropical countries.

(*l*) Compose a circular letter from the manager to all employees informing them of changes in the routine of clocking in and out.

(*m*) As manager of a small light-engineering factory write to an engineering consultant to arrange for him to test one of your products.

(*n*) A research laboratory has asked you, as a small manufacturer of special steel parts, to make for experimental use a special long narrow-bored tube. Write a polite letter explaining that the job is an impossible one, particularly if they require the accuracy they state.

(*o*) As personnel officer in a large organization, draft a circular letter to shop stewards explaining in brief the advantages of a change from individual to bay or section bonus schemes and calling a meeting for discussion.

(*p*) Write a letter confirming a telephone call you have made urgently requesting a replacement part for a key machine. Confirm particularly the type, number, etc., of the part required.

(*q*) Write a letter of resignation from your firm.

(*r*) As site manager of a building project, draft a tactful letter to the Local Authority who are complaining about muddied and greasy roads near the site explaining that (i) other contractors are also to blame, and (ii) what you propose to do about it.

(*s*) There have been rumours of discontent and exploitation in your factory. As public relations officer, draft a letter to the local Press carefully demolishing these rumours.

REPORTS

A report is usually a statement of facts presented for the attention of someone seeking information. The aim is to present the information in a clear and usable form, and the report may vary from a small cyclostyled leaflet to a lengthy printed volume, complete with index, appendices, etc.

Routine or recurring reports are now mainly made by filling in specially designed forms; but the Special Report needs to be carefully prepared, written, and edited.

The Terms of Reference

These are the instructions that you are given to help you prepare and write the report. They must make clear the following conditions:

(*a*) The kind of information required.
(*b*) How much you should provide.
(*c*) How urgent the report is.

The terms of reference of a report might read as follows:

Please investigate the possibility of replacing the machinery in workshop A, and recommend the most economical method of operating the change-over. A brief report is required before the joint consultation meeting on 15th February.

Preparing a Report

In collecting material for a report you must keep strictly to the terms of reference. You must have a plan of action setting out the lines of enquiry and your sources of information; this is particularly important if you have to collect the material within a specified time.

In collecting material for a report you should follow three stages:

(*a*) Discover what you need.
(*b*) Analyse what you already have.
(*c*) Obtain what else is necessary.

Material may be obtained by reading, interviewing people, circulating questionnaires or by direct observation and experiment.

Arranging the Material

When you have collected all the available information you select those items which come within the scope of your Terms of Reference and arrange the material in a definite order. This may be:

Chronological—following the stages of the investigation.
Geographical—under the headings of different areas.
Increasing complexity—starting with the well-known information and progressing to the new material.
Functional—separating the information into the different functions—*e.g.*, reports on machinery and processes
By rejection—showing unsuitability of other methods.

There are other ways of arranging the material. The choice of method depends on the individual report and the potential reader.

Drafting the Report

This must be done within a framework. The basic outline of a short report is:

Terms of Reference
Facts
Conclusions
Recommendations

You must first restate your terms of reference. Then you must present the facts in a logical order, and from those facts you will usually be required to draw certain conclusions and make recommendations.

The draft report is composed under these headings in the following manner:

Terms of Reference
A restatement of the original terms of reference.

Facts

This section should contain a clear statement of all the facts relevant and necessary to this report.

Opinions and generalizations which may mislead are not required.

Details and figures which may confuse should be relegated to an appendix.

Conclusions

What you consider the facts indicate.
What you think is likely to happen.
What you consider is being done.

Recommendations

What course of action should be taken either now or in the future.

The Final Report

This may take any of a number of forms of presentation.

(*a*) *Letter report:* usually submitted after a personal investigation. Written in the first person but with a definite plan shown

E.g. Introductory paragraph
Paragraphs showing the findings and facts
A paragraph giving the conclusions
A final paragraph with recommendations

There will not necessarily be sub-headings within the letter form, but the paragraphs may be numbered. Marginal sub-headings or headings may be used.

(*b*) *Memorandum:* report written on a specially headed memorandum form with no salutation or complimentary close. Otherwise it is similar in design to the letter report.

The memo. is used mainly for transmission of information and instructions within an organization.

(*c*) *Formal report:* cyclostyled or printed presentation, with title, headings, sub-headings, etc., formally displayed and tabulated.

Variation in Arrangement of Report

For added emphasis and effect, or to deal with particular situations, there are different ways of arranging the material.

The following are suggestions which may be used in the report writing exercises at the end of this section:

1. Terms of reference
   Previous findings
   Present investigation
   Facts
   Conclusions
   Recommendations

2. Terms of reference
   Survey of present system
   Investigation
   Results
   Suggested remedy

3. Terms of reference
   Present method
   Comment on present method
   Comparison with other methods
   Suggested new method

4. Summary of the conclusions and recommendations
   Terms of reference
   Investigation
   Results
   Details of conclusions and recommendations
   (Known as the 'Inverted Report.' Useful for non-specialist reader)

5. Terms of reference
   Present state of enquiry
   Progress
   Future activities
   (Known as an 'interim report')

You will be able to devise many more variations on the basic structure. Select the arrangement according to the reader, the time in which he will be able to read and act, the importance of certain parts of the report, and the nature of the information you are providing.

Editing the Report

Always read through your work, whether in the rough draft stage or the final manuscript or typescript. Be a ruthless self-critic. Do not be afraid to weed out awkward phrases or semi-relevant material.

Exercises

1. The following vague instructions are for special reports. Construct what you consider to be reasonable terms of reference for each:

   (*a*) You remember that fire in the paint shop the other week? Let me have a report on it, will you?

(*b*) Work out what strain we can expect on the arches of that bridge.
(*c*) You back-room boys should be able to work up a new method of dyeing these artificial fibres. Work something out and let me know.
(*d*) What do the blokes in the shop think about this new machinery?
(*e*) Some people think we ought to open up a new branch at Yarmouth. What do you think?
(*f*) If I don't find out what problems there are in re-wiring the office block I shall be in hot water at next week's meeting.
(*g*) I don't know if we should embark on a new sales campaign at this stage. What do you advise?
(*h*) Why have there been so many accidents in the machine shop recently? I must know quickly.
(*i*) The employees are grumbling about rumours of redundancy. How did the rumour get around?
(*j*) Do we really need a new sports pavilion?

2. Provide in outline form, *i.e.* in heads and sub-heads only with a reference scheme, a report on the following subjects:

(*a*) A progress report on alterations to a workshop.
(*b*) A report on the testing of a new piece of machinery.
(*c*) A report on the comparative sales efficiency of certain districts.
(*d*) The description of a new method of dust control in the factory.
(*e*) A report on newly-discovered methods of three-dimensional milling.
(*f*) A report on the possibility of using new telemetry techniques in rocket testing.
(*g*) A memorandum setting out suggestions for the improvement of office facilities.
(*h*) A report suggesting a re-organization of the communication system in the factory.
(*i*) A report on the investigations into structural defects in an airframe.
(*j*) A memorandum indicating possible new methods of job assessment.

3. Compose reports in accordance with the following instructions. Each report should be between 200 and 500 words in length, and consideration should be given to plan, style of presentation, setting-out, etc., rather than to the content alone.

Some advice will be given concerning the outline the report might take.

(*a*) Please investigate the present method of introducing apprentices to their work with the Company, and suggest, in a personal letter report, any means of improving the system.

(*b*) You have been asked to suggest a suitable design of safety valve for a new type of coke boiler which works at high pressure to provide heating for a large workshop. Make your report and recommendations in the form of a memorandum.

(*c*) Please test the new lathe in the workshop and make a report (consider particularly—tests, equipment, results, observations, recommendations). Submit it in the form of a memorandum.

(*d*) Examine the arrangements for the maintenance of machinery in the workshop and suggest, in a personal letter report, any necessary amendments or additions to the present routine.

(*e*) You have tested a new 700 cm^3 economy four-seater car. Compose and display a concise Road Report on the car.

(*f*) Adjoining your factory is an old warehouse which is the property of the firm. The management has decided to make use of the building either as a canteen or a new workshop, and has asked you as works manager to investigate. Submit a short report, paying particular attention to state of building, accessibility, main services, priority of needs in the organization, possibility of conversion, rough estimate of cost, etc.

(*g*) There is the possibility of changing from belt-drive to individual motor drive in the machine shop. Please submit a report setting out the facts about each method. You are not required to comment or to recommend, but merely to submit a memorandum with the facts.

(*h*) Local transport difficulties have resulted in a request from the local authority for your organization to stagger the hours of work. Submit a report on the possible methods of doing this, pointing out any difficulties and paying attention to problems of assessing overtime, difference between office and shop staff, union reaction, production difficulties, etc. Present this as a formal report.

(*i*) Please investigate the recent accidents and fires caused by electricity in the factory, paying particular attention to their origin and to means of prevention. Submit this in the form of a memorandum.

(*j*) You have been asked to comment on a suggested local advertising plan involving use of hoardings, distribution of handbills and small posters, posting of circular letters, etc.

Make your report, in memorandum form, paying particular attention to possible local reaction.

(*k*) The heating system in your office or workshop is old and inefficient. Submit a formal report on the possible systems that might be used, and recommend a replacement.

(*l*) The management has decided to launch a safety-poster campaign, and you have been asked to make suggestions. Submit a short memorandum on the notice-spaces available, the possible reactions by the employees, and balance these considerations against the expense involved. Draw your conclusions and make recommendations.

(*m*) A new bonus and incentive scheme, only recently introduced, has aroused much criticism from the employees. You have to investigate the causes of complaint and recommend some action. Use the personal letter form for this report.

(*n*) The management is considering the installation of a loudspeaker system in the factory. You have been asked to give your advice regarding possible staff reaction; suitable siting of loudspeakers; limits upon use of the system; estimated interference with production, etc. Submit this in memorandum form.

(*o*) The drawing office needs modernising and among the problems to be faced is that of suitable lighting. You have been asked to report upon suitable lighting systems, problems of glare and strength of lighting, etc., and should submit your findings and recommendations in a formal report.

(*p*) Noise has become a disturbing element in the life of the factory. The noise of machinery inside has been supplemented by drills and bull-dozers busy on demolition and construction nearby. You have been instructed to assess the nuisance value of this noise, by using a noise-meter, tape recorder, etc. Make your report, and any recommendations.

(*q*) Submit a brief report on a hardness test which you have carried out for another department.

(*r*) There is evidence that one section of the production team is not pulling its weight. You are asked, as a methods study man, to investigate and to submit a personal and confidential report of the facts, accompanied by your recommendations.

(*s*) The tea-break is proving a problem and is causing a disproportionate drop in production. You are asked to investigate and to report on possible means of providing refreshment without disturbing production. You should pay particular heed to automatic dispensing systems; allowance of time in lieu; trolley service; canteen facilities; staggered breaks. Present the material in a suitably displayed and headed formal report.

(*t*) Your research section has just finished designing a new electronic testing device. You have been asked by the Production Planning department to report on progress in the development and to make clear what advantages the new device possesses. Submit a memorandum with this information, together with an estimate of production difficulties.

V

COMPREHENSION AND INTERPRETATION

Understanding what you read is not an automatic process; it involves hard work and demands intense concentration, particularly if the writer is not helpful in his presentation of the material. You must be able to *read*, and by that I mean able to take the ideas from the page with speed and accuracy. This also demands concentration, and most of us could read with far greater efficiency than we do. The average person is said to read about 250 words per minute with roughly 70 per cent comprehension—not a high performance.

READING EFFICIENCY

This depends upon:

(*a*) *Good reading ability:* Being able to read in word groups rather than one word at a time. Silent reading should be *silent*, and you must avoid the habit of 'vocalizing' words or thinking of them as 'sound' units.

(*b*) *Concentration:* Slow reading can produce listlessness, lack of concentration and greater chances of distraction. To read quickly and efficiently means that you can see the overall framework of your reading and assimilate the material more easily.

(*c*) *Sound background knowledge:* The more you know of a subject, the more easily will you understand anything associated with it.

(*d*) *Wide vocabulary:* The greater the range of your vocabulary the quicker you can grasp the meaning of a passage and the fewer interruptions of your reading.

(*e*) *A knowledge of the English language:* A sound grammatical knowledge means fluent reading.

(*f*) *A knowledge of what you are looking for:* Aimless reading is usually a waste of time. Reading should have a purpose.

READING TO UNDERSTAND

Be methodical in dealing with a passage of reading material. Use the following method:

1. Skim the passage for several seconds, looking for:
   (*a*) an idea of its difficulty
   (*b*) its subject and the way the subject is developed
   (*c*) the structure and development of paragraphs
2. Read the passage in more detail examining the paragraphs for their topic sentences and development.
3. Store in your mind the information you require from each paragraph and section.

If the passage is well written, your task should not be a difficult one. Your skimming will give an idea of the general paragraph structure; then, as you read, you can begin to break down the paragraphs into their main ideas, following the flow of thought within each paragraph and the movement from one paragraph to another.

If it is difficult to pick out the topic sentence, you can make a mental telegraphic version of each paragraph by picking out the essential words:

E.g. When two metallic surfaces are apparently in contact, the application of a small amount of lubricant causes a very thin film to be formed on each of the surfaces. This thin film adheres to the surface and is said to be an *adsorbed* film, and it is now recognized that such a film is really in chemical union with the surface. (61 words)

This can be reduced as follows:

Two metallic surfaces contact, application of small amount lubricant causes thin *adsorbed* film on surfaces and chemical union with surface. (20 words)

Develop, by practice, the ability to reduce the material to its elements.

EXTRACTING THE INFORMATION

This is not merely a matter of soaking up all the information in the passage you are reading; it is a process of judgement and criticism in which you, as reader, are involved as much as the writer.

When you set out to read a report or a book there is usually a purpose involved: to gain information, to extend your knowledge, to help you to pass an examination, to enable you to decide upon a course of action, for sheer curiosity, or for entertainment. Know your purpose in reading, and be quite clear about the material you will need to memorize and recall, and that which can be merely noted for future reference.

INTERPRETATION

If you have read a passage and have taken from it the facts you require, that may be sufficient for your purpose; but there are times when you are required to go a step beyond comprehension to the technique of interpreting. This involves expressing your understanding in words of your own choosing, or in words that are selected to suit a fresh audience. One example of this is the need to simplify a set of instructions for a particular section of non-specialist workmen.

Remember that it is the process of *expressing in your own words*, and not just the action of changing individual words into the first synonym that comes into your head. It is re-thinking a passage and saying to yourself—and to your audience—"It seems to me that what the writer means is. . . ." Remember, also, that your expression must be clear and concise and written in good English.

Exercises in Comprehension and Interpretation

Paragraph Analysis

Exercises

1. Indicate the main topic of each paragraph, then reduce the paragraph to telegraphic form.

(*a*) When in its pure state, iron is of little practical use and the difficulty involved in obtaining it from the ore is not worth the trouble. Pure iron is so soft that machining results in a rough finish and in bad tearing of the metal. (45 words)

(*b*) Magnetic poles are always in pairs—the north pole at one end of the magnet and the south pole at the other. You cannot break a magnet in half and remove its north pole, for all that will happen is that a further south pole will arise at the broken end, and you will be left with two smaller magnets each with its opposite poles. (65 words)

(*c*) Industrial concerns are involved in a great deal of expense as a result of the action of rust upon equipment and materials, particularly when they are not in use. Objects corrode quickly when in store or when being transported, and cleaning these objects, or replacing them when the action of corrosion has gone too far, is expensive in both money and time. (62 words)

(*d*) The coleopter is a type of aircraft capable of vertical take-off at high speeds, and of supersonic flight at high altitudes. To achieve this it uses jet propulsion within an annular wing—an inside turbo-jet for initial propulsion and control, with the wing later forming the outer wall of a ram jet to give the high speeds at higher altitude. (63 words)

(*e*) Though it is usual to divide materials into two categories, conductors and insulators, it should be realized that these are only relative terms. No material is a perfect conductor and no material is a perfect insulator. In general, metals such as copper and silver are very good conductors, whereas non-metallic materials such as rubber and glass are good insulators. (60 words)

(*f*) The calculus deals with quantities in the process of growth and with rates of growth. It involves also the classification of all quantities into two classes—constants and variables. Constants, quantities of fixed value, are usually represented by letters from the beginning of the alphabet, and variables, quantities capable of growth, by letters towards the end of the alphabet. (60 words)

(*g*) One of the most remarkable features of our power of vision is its ability to adapt itself to the wide range of brightness in our surroundings. When we pass from the bright glare of sunshine in the open air to the dimly lit interior of a building, the intensity of illumination may fall by a factor of a million to one. If the eye is embarrassed by such a drop in intensity it is only for a little while. Soon it adapts itself to the feeble illumination. After a long rest in the dark, objects that were invisible at first may appear insupportably bright.

Under favourable conditions the sensitivity of the human eye is indeed amazing. (117 words)

Read each of the following paragraphs quickly and carefully, trying to get complete understanding at one reading. If this does not succeed, give a second skim through the paragraph. Then cover the paragraph and without further reference to it answer the questions set. Word-for-word accuracy is not essential.

1. No measurement can be made with absolute accuracy. We can only determine the limits between which the accurate result would lie. If we measure the length of an iron bar to the nearest millimetre and the result is 186, to three significant figures, the true length of that bar is less than 186.5 and greater than 185.5.

(*a*) Write down the topic sentence of the paragraph.
(*b*) State the example given of determining the limits between which the accurate result would lie.

2. Mechanics is related to everyday experience. As the study of forces it deals with most objects, because even when something is at rest forces are still acting. A small child pushing against a table exercises a force even though he may not move it, while the same force exerted by an adult may move the table several feet. Even the chair on which you are sitting is exerting an upward force to balance your weight. If it did not you would fall through it to the floor.

(*a*) Write down the topic sentence.
(*b*) How many examples are given of forces acting on something at rest?
(*c*) Quote these examples.

3. Electrolysis is the decomposition of a substance called an electrolyte by passing through it a direct current between two electrodes, an anode and a cathode. The substance in the electrolyte splits up into two sorts of electrically charged ions which are attracted to the electrodes where they may be freed or may act chemically upon the electrodes. The mass of ion liberated is proportional to the quantity of electricity passed through the electrolyte and to the chemical equivalent of the ion.

(*a*) Write down the topic sentence.
(*b*) What two courses of action are open to the ions on being attracted to the electrodes?
(*c*) Faraday's law concerning the mass of an ion is mentioned. Quote it.

4. A Geiger counter is an instrument used for the detection of radio-activity. It is basically a metal cylinder containing a conducting wire from which it is insulated by a vapour; the cylinder and wire are maintained at a potential difference just below that required for a current to flow through the vapour. If a nearby substance gives off cosmic or alpha rays these will make the vapour conductive and a current will pass through the vapour. This can be measured, recorded, and counted.

(*a*) Write down the topic sentence.
(*b*) Describe the basic structure of the counter.
(*c*) What rays affect the conductivity of the vapour?

5. The heat in a substance is associated with the motion of its molecules. The hotter the substance the more vigorous are the vibration and motion of its molecules. If heat is applied continuously to a solid it relaxes the cohesion of the molecules and a point is reached at which the vibration of the molecules is such that the solid changes into liquid in which the molecules can move about more freely. On further addition of heat to the liquid the motion of the molecules is increased still more, and a point is reached at which the liquid begins to change into vapour and then becomes gas. The heat applied dissociates the molecules of liquid from one another so that they fly apart and remain separate in the gaseous state. Heat in a substance may, therefore, be regarded as molecular energy. Radiant heat, however, which travels with the velocity of light, does not require a material medium for its transmission.

Heat Engines, by A. C. Walshaw

(*a*) Write down the topic sentence.
(*b*) What is meant in the paragraph by the terms cohesion, dissociates, medium?
(*c*) Explain in brief terms the stages in increased molecular motion.
(*d*) Why is radiant heat different from other forms of heat?

General Exercises

The following exercises are designed to test your reading, comprehension and interpretation. At the beginning of each passage is a brief note of the material you will read and an idea of what you are to look for.

Read each passage carefully, then move on to the questions. The first two excerpts will be followed by a straightforward set of questions and a practical exercise, then the exercises will take the following form:

Section I a series of 'snap' questions designed to test your initial grasp of the material. These questions should be answered without reference back to the passage.

Section II questions asking for more detailed information and interpretation.

Section III further exercises, usually practical, and suitable for both individual and group work.

N.B. If you are unable to answer the questions in Section I without help, go back to the passage and make sure that you understand it more thoroughly, then move to the questions in Section II.

1. Many students of different background and different ability will find employment in a machine shop. The following extract from a careers booklet on engineering gives a brief survey of the range of jobs.

The Machine Shop

Whatever other processes may be used for them, most parts are machined at some stage, and the machine shop usually employs the largest number of workers. Almost all castings and forgings need to be machined and many parts are made entirely from rods or bars of metal by machining them.

Where large numbers of each part have to be machined, the work is made as simple as possible for the operatives, so that they can work quickly and easily, yet still turn out work of high quality. For example, where parts have to be machined one after another from the end of a long bar of metal, there is usually a 'stop,' which makes it easy to move the right length of fresh metal into position each time. Stops are also fitted to prevent a cutting tool from being fed too far into the metal. Where a partly shaped piece of metal is to be machined, there will be a 'fixture' to hold it in position, so designed that the part cannot be put in the wrong way round. In drilling and similar operations, 'jigs' are often used; these hold the part firmly and have a guide hole for each hole to be machined in the work piece; the operator has only to bring the guide holes in turn under the drill of his machine to make sure of drilling in exactly the right place.

Some of the machines are so made that, once prepared for a run of work, all they need is to be kept supplied with metal. One or two 'machine minders' may look after a score of such machines, putting long bars of metal into them from time to time; the machines will go on running for hours, turning out parts in great quantities. Other machines call for a real knack in fitting the part quickly and securely in place, and the operator can make a big difference to the finish of the work according to the care he takes in operating the controls steadily and at the correct rate. Sometimes operators may be taught to 'set up' their machines themselves, adjusting the fixtures, stops, and working speed for each batch of work. There is employment in the machine shop, therefore, for boys of different ability and outlook.

Choice of Careers—Engineering, H.M.S.O.

(*a*) Why is the machine shop such an important and extensive section of a factory?

(*b*) What is the meaning in the passage of the following terms:
machining; operatives; 'stop'; 'jigs'; knack

(*c*) Why does a fixture have to be carefully designed?

(*d*) What type of machine work can be left to a 'machine minder'?

(*e*) When can the machine operator make a big difference to the work he is doing?

(*f*) What is meant by 'setting up' a machine?

PRACTICAL EXERCISE

Draw an explanatory chart showing the various types of work. Sketch the machines, if you wish to, and give a brief caption indicating the amount of skill and responsibility required.

2. In animals the ability to fly depends upon wings, whether they be the great span of the abatross or the loose skin of a flying squirrel. Wings are limbs able to produce forces that will counteract gravity and move a body through the air. The following extract from a book on animal movement describes the basic action of a wing. Read it carefully for the information it provides.

In its simplest form a wing can be thought of as a flat plate sticking out sideways from the body of a bird or from the fuselage of an aeroplane. If this flat plate be held edgewise in a current of air, the force of the current will tend to drag the plate with it, downstream; and only an external force tending to pull it upstream can hold it in position. If, however, we slightly raise the edge that faces upstream (or lower the downstream edge) the plate will tend to rise bodily in the air, although, at the same time, the force tending to drag it along downstream increases. The moving air is thus exerting two forces on the inclined plate: (*i*) a *lift* force tending to raise the plate; and (*ii*) a *drag* force pushing it back in the direction of the air stream. True gliding flight becomes possible when these two forces are so adjusted as to be exactly equal to the weight of the bird or aeroplane.

Tests with various sizes and shapes of wings, in winds of various speeds, have shown that the lifting effect of the air depends on four main factors: the shape of the wing; its size; the angle at which it is inclined to the direction of the wind; and the speed at which the wind is travelling past it.

The mechanical efficiency of a wing depends largely on its ability to develop a large lifting force for a relatively small increase in backward drag; but, from a practical point of view, the ability to fly depends on the extent to which the exact balance can be maintained between the lift and drag forces of the air on the one hand, and the pull of gravity on the other. Stable, continuous flight can only happen when the wings are so arranged that any slight accidental disturbance of the air-flow is automatically compensated.

How Animals Move, by James Gray

(*a*) What is the effect of a current of air upon a wing that is directly edgewise to it?

(*b*) What difference is made by slightly raising the forward edge of the wing?

(*c*) Write a definition of (i) lift and (ii) drag.

(*d*) Under what state of these two forces does true gliding flight become possible?

(*e*) The lifting effect of the air on a wing depends upon four main factors. Name them.

(*f*) Explain in your own words the practical problem of flying as described in the last paragraph.

PRACTICAL EXERCISE

Draw diagrams of wing sections to show the points explained in the excerpt. Show clearly the effects of lift and drag.

3. In Sir John Hunt's book *The Ascent of Everest*, there is a vivid description of the final conquest of the peak. The story is told by Sir Edmund Hillary, and the following extract begins at the formidable rock step that met Hillary and Tenzing on the last stage of their climb. Read this passage more slowly than usual, taking in the full atmosphere of that tense part of a great endeavour.

After an hour's steady going we reached the foot of the most formidable-looking problem on the ridge—a rock step some forty feet high. We had known of the existence of this step from aerial photographs and had also seen it through our binoculars from Thyangboche. We realized that at this altitude it might well spell the difference between success and failure. The rock itself, smooth and almost holdless, might have been an interesting Sunday afternoon problem to a group of expert rock climbers in the Lake District, but here it was a barrier beyond our feeble strength to overcome. I could see no way of turning it on the steep rock bluff on the west, but fortunately another possibility of tackling it still remained. On its east side was another great cornice, and running up the full forty feet of the step was a narrow crack between the cornice and the rock. Leaving Tenzing to belay me as best he could, I jammed my way into this crack, then kicking back with my crampons I sank their spikes deep into the frozen snow behind me and levered myself off the ground. Taking advantage of every little rock hold and all the force of knee, shoulder, and arms I could muster, I literally cramponed backwards up the crack, with a fervent prayer that the cornice would remain attached to the rock. Despite the considerable effort involved, my progress although slow was steady, and as Tenzing paid out the rope I inched my way upwards until I could finally reach over the top of the rock and drag myself out of the crack on to a wide ledge. For a few moments I lay regaining my breath and for the first time really felt the fierce determination that nothing now could stop us reaching the top. I took a firm stance on the ledge and signalled to Tenzing to come on up. As I heaved hard on the rope Tenzing wriggled his way up the crack and finally collapsed exhausted at the top like a giant fish when it has just been hauled from the sea after a terrible struggle.

I checked both our oxygen sets and roughly calculated our flow rates. Everything seemed to be going well. Probably owing to the strain imposed on him by the trouble with his oxygen set, Tenzing had been moving rather slowly but he was climbing safely, and this was the major consideration. His only comment on my enquiring of his condition was to smile and wave along the ridge. We were going so well at three litres per minute that I was determined now if necessary to cut down our flow rate to two litres per minute if the extra endurance was required.

The ridge continued as before. Giant cornices on the right, steep rock slopes on the left. I went on cutting steps on the narrow strip of snow. The ridge curved away to the right and we had no idea where the top was. As I cut around the back of one hump, another higher one would swing into view. Time was passing and the ridge seemed never-ending. In one place, where the angle of the ridge had eased off, I tried cramponing without cutting steps, hoping this would save time, but I quickly realized that our margin of safety on these steep slopes at this altitude was too small, so I went on step-cutting. I was beginning to tire a little now. I had been cutting steps continually for two hours and Tenzing, too, was moving very slowly. As I chipped steps around still another corner, I wondered rather dully just how long we could keep it up. Our original zest had now quite gone and it was turning more into a grim struggle. I then realized that the ridge ahead, instead of still monotonously rising, now dropped sharply away, and far below I could see the North Col and the Rongbuk glacier. I looked upwards to see a narrow snow ridge running up to a snowy summit. A few more whacks of the ice-axe in the firm snow and we stood on top.

Section I

(*a*) How did the climbers know of the difficulty presented by the rock step?
(*b*) What means did Hillary find of overcoming this barrier?
(*c*) What particular scientific aid was being used by the climbers at that altitude?
(*d*) Why had the climbers no idea of their distance from the summit?
(*e*) What activity began to exhaust the two climbers during the last stages?

Section II

(*a*) Why did Hillary consider the rock step to "spell the difference between success and failure"?
(*b*) Why do you think Hillary was filled with "fierce determination" after his climb up the crack between the rock and the cornice?
(*c*) Why was Tenzing moving slowly?

(*d*) What conditions began to dishearten the climbers?

(*e*) Give two examples from the passage of Sir Edmund Hillary's sense of judgement and lack of foolhardiness.

Section III

Refer to the book *The Ascent of Everest*, and from it compile a brief list of the technical equipment used in the expedition—particularly the equipment made possible by recent scientific advances. Against this list, make a note of the human qualities needed by a great climber, then discuss in your group the following questions:

(*a*) Has the spirit of adventure encouraged by adversity disappeared from mountain climbing?

(*b*) Is the use of much scientific equipment tantamount to cheating?

(*c*) In what sense is climbing an achievement for (i) an individual, (ii) a team?

(*d*) Is it true to say that "Adventure is a series of brief exciting moments interspersed with long stretches of fatigue and boredom"?

(*e*) Is there any purpose in climbing mountains? Does it further our scientific knowledge or improve our character?

4. In the spring of 1945, the scientists at Los Alamos were at work upon a dangerous project—the assembling of the atomic bomb. Amongst them was a Canadian physicist, Louis Slotin—a man keen upon excitement and adventure, who had served in the Spanish Civil War as an anti-aircraft gunner and who had volunteered for the Royal Air Force immediately on the outbreak of the Second World War.

Slotin was a highly qualified biophysicist from London University and from being a fighter pilot he became a scientist active in his task of mounting the internal mechanism of the Alamogordo experimental bomb, and then in the preparations for the Bikini atoll test.

The following extract from Dr Robert Jungk's *Brighter than 1000 Suns* tells the tragic story of Slotin's work and dispels some of the illusions about the academic calm of scientific research.

. . . Slotin had been busy testing the interior mechanism of the experimental bomb. It consisted of two hemispheres which would come together at the moment of release, thereby enabling the uranium they contained to unite in a 'critical mass.' The determination of this 'critical size'—referred to simply as 'crit' in the Los Alamos jargon—had been one of the chief problems studied by the theoretical department. But the quantity of uranium required, the scattering angle and range of the neutrons to be emitted by the chain-reaction, the speed at which the two hemispheres would have to collide, and a whole series of other data could only be approximately estimated. If absolute precision and certainty were to be attained, it could only be by way of experiment in every case. Such experiments were assigned to the group in charge of O. R. Frisch, the discoverer of fission, who had been brought to Los Alamos from England. Slotin was one of the members of this group. He was in the habit of experimenting without taking any special protective measures. His only instruments were two screwdrivers, by means of which he allowed the two hemispheres to slide towards each other on a rod, while he watched them with incessant concentration. His object was to do no more than just reach the 'critical point,' the very first step in the chain-reaction, which would immediately stop the moment he parted the spheres again. If he passed the point or was not quite quick enough in breaking contact, the mass might become 'super-critical' and produce a nuclear explosion. O. R. Frisch himself had once nearly lost his life during one of these experiments at Los Alamos.

Slotin knew, of course, by how narrow a margin his chief had escaped death. But the daring young scientist thoroughly enjoyed risking his life in this way. He called it 'twisting the dragon's tail. . . .'

On the 21st May, 1946, not quite a year later, Slotin was carrying out an experiment, similar to those he had so often successfully performed in the past. It was connected with the preparation of the second atom-bomb test to be performed in the waters of the South Sea atoll of Bikini. Suddenly his screwdriver slipped. The hemispheres came too close together and the material became 'critical.' The whole room was instantly filled with a dazzling, bluish glare. Slotin, instead of ducking and thereby possibly saving himself, tore the two hemispheres apart with his hands and thus interrupted the chain-reaction. By this action he saved the lives of the seven other persons in the room. He had realized at once that he himself would be bound to succumb to the effects of the excessive radiation dose which he had absorbed. But he did not lose his self-control for a moment.

He told his colleagues to go and stand exactly where they had been at the instant of the disaster. He then drew on the blackboard with his own hand, an accurate sketch of their relative positions, so that the doctors could ascertain the degree of radiation to which each of those present had been exposed.

As he was sitting with Al Graves, the scientist who had been most severely affected, apart from himself, by the radiation, waiting at the roadside for the car which had been sent for to take them to hospital, he said quietly to Graves: "You'll come through all right. But I haven't the faintest chance myself." It was only too true. Nine days later the man who had experimentally determined the 'critical mass' for the first atom-bomb died in terrible agony.

Section I

(*a*) What was Louis Slotin's job at Los Alamos?
(*b*) What were the only instruments he used?
(*c*) What was his second experiment connected with?
(*d*) How was the accident caused?
(*e*) What was the cause of Slotin's death?

Section II

(*a*) What was meant by 'critical mass'?
(*b*) What factors had to be decided experimentally?
(*c*) What were the dangers in Slotin's experiment?
(*d*) Describe how Slotin sacrified his life to save the other scientists.
(*e*) What great presence of mind did Slotin show after the accident?
(*f*) Describe what makes a nuclear accident different from other fatal accidents involving explosives.

Section III

Since Louis Slotin's tragic experiment, and the even greater tragedy of Hiroshima and Nagasaki, the development of the Hydrogen bomb has presented an even more appaling danger to mankind.

Prepare a chart showing how the nuclear bomb has developed and how its potential effect has increased. Against that information display the progress made in nuclear benefits—atomic power stations, atomic ships, radio-active isotopes, in medicine, etc.

DISCUSSION TOPICS

1. Are the benefits of nuclear physics inseparable from the great dangers?
2. Have the scientists a moral responsibility for the tragedy and fear caused by nuclear bombs?

5. The following extract is from Sir Francis Galton's famous work *Inquiries into Human Faculty and its Development.* The book was first published as long ago as 1883, and marked a stage in the progress of experiment which was rapidly broadening the sciences of psychology and anthropology. This short section deals with Galton's well-known experiment with whistles to test the audibility of shrill notes.

I contrived a small whistle for conveniently ascertaining the upper limits of audible sound in different persons. . . . I made a very small whistle from a brass tube whose internal diameter was less than one-tenth of an inch in diameter. A plug was fitted into the lower end of the tube, which could be pulled out or pushed in as much as desired, thereby causing the length of the bore of the whistle to be varied at will. When the bore is long the note is low; when short it is high. The plug was graduated, so that the precise note produced by the whistle could be determined by reading off the graduations and referring to a table.

On testing different persons, I found that there was a remarkable falling off in the power of hearing high notes as age advanced. The persons themselves were quite unconscious of their deficiency so long as their sense of hearing low notes remained unimpaired. . . . They are indignant at being thought deficient in the power of hearing, yet the experiment quickly shows that they are absolutely deaf to shrill notes which the younger persons hear acutely, and they commonly betray much dislike at the discovery. . . .

I have tried experiments with all kinds of animals on their powers of hearing shrill notes. I have gone through the whole of the Zoological Gardens, using an apparatus arranged for the purpose. It consists of one of my little whistles at the end of a walking stick—that is in reality a long tube; it has a bit of india-rubber pipe under the handle, a sudden squeeze upon which forces a little air into the whistle and causes it to sound. I hold it as near as is safe to the ears of the animals, and when they are quite accustomed to its presence, and heedless of it, I make it sound; then if they prick their ears it shows that they hear the whistle; if they do not it is probably inaudible to them. Still it is very possible that in some cases they hear but do not heed the sound. Of all creatures I have found none superior to cats in the power of hearing shrill sounds. . . . Cats, of course, have to deal in the dark with mice, and to find them out by their squealing. Many people cannot hear the shrill squeal of a mouse. . . . Cats are differentiated by natural selection until they have a power of hearing all the high notes made by mice and other little creatures that they have to catch. . . . Small dogs also hear very shrill notes, but large ones do not. I have walked through the streets of a town with an instrument like that which I used in the Zoological Gardens, and made nearly all the little dogs turn round, but not the large ones. . . . Ponies are sometimes able to hear very high notes. I once frightened a pony with one of these whistles in the middle of a large field. My attempts on insect hearing have been failures.

Section I

(*a*) What was Sir Francis Galton's purpose in constructing the special whistle?
(*b*) What did he discover from his experiments with human hearing?
(*c*) Why did he construct a special tube for the apparatus he used at the Zoo?
(*d*) Which animal did he find to be the superior in hearing shrill sounds?
(*e*) What results did he get with dogs?

Section II

(*a*) Describe carefully the first whistle that Galton constructed. Make a sketch if it helps.
(*b*) Why do you think people were annoyed at being told they could not hear the notes?
(*c*) Describe the additions made to the whistle apparatus for use at the Zoo.
(*d*) Why do cats hear shrill sounds well?
(*e*) Explain the meaning and use in the passage of each of the following terms:

contrived in "I contrived a small whistle"
graduated in "The plug was graduated"

unimpaired in "hearing low notes remained unimpaired"
differentiated in "cats are differentiated"
natural selection in "differentiated by natural selection"

Section III

Collect the results that Sir Francis Galton has obtained and note them in a neat list. Then refer to the appropriate reference work in your college library and list the up-to-date information about human and animal hearing. Compare the two sets of results and answer these questions:

(*a*) How near was Sir Francis to the 'truth'?
(*b*) What important factors were unknown to him?
(*c*) How 'scientific' was his method?
(*d*) Why were his experiments with insects a failure?

6. The history of measurement, particularly the measurement of time, is a long story and one closely connected with the stars. Read this excerpt from L. Hogben's *Mathematics for the Million* in which is described the geometrical knowledge acquired by the ancients to meet their practical needs. Concentrate on the geometry as you will be asked to explain aspects of the measuring systems and to provide diagrams.

The need for accurate measurement grew naturally from the practice of time-keeping, which was an essential prerequisite of settled city life. It is fairly certain that people began to measure angles long before they troubled themselves very much with measuring lengths. The number of days in a year had been fixed by the heliacal rising of Sirius at the very beginning of settled city life along the Nile. Watching for the rising of a particular star or constellation involves knowing at what point on the horizon it will appear, and we have ample evidence that Neolithic man had learned to set up crude monuments to fix the direction of celestial phenomena, before there were cities which have left any permanent remains. To recognize the direction in which an object will be seen on the horizon we must observe it from some fixed point, and be able to refer to some fixed line. Two fundamental lines of reference, the prime meridian which joins the north and south points on the horizon, and the line at right angles to it joining the east and west points, had probably been settled before the great calendar civilizations began. Their discovery constitutes the first mathematical problem in the social practice of man.

The choice of the north-south meridian was based on the position of the sun's shadow when it is shortest at midday. It also points directly towards the place in the heavens around which the stars revolve by night. At the time when the Pyramids were built, a bright star in the constellation of Draco revolved in a tiny circle (of three degrees) round it. The position of the constellations changes in the course of centuries, and today the celestial pole is recognized by the star Polaris, our Pole star, which is only one degree off. We have a record of the way in which the exact location of the noon shadow was fixed in ancient times. In the sand or soft earth around the shadow pole or obelisk, which was the public clock, a circle was traced with a piece of cord. The two points where the shadow just touched the circle were marked and the angle between these two points and the shadow pole was bisected first, perhaps, by stretching a cord between them and folding it in halves, later by tracing arcs of equal radius about the two points.

The recognition of the east and west meridian or *equinoctial* presented a similar problem, and the orientation of burial sites shows that it also antedates any city life. We are so accustomed to be taught that the sun rises in the east and sets in the west that many of us do not realize that it only does so on two days of the year—the vernal and autumnal equinoxes, when night and day are of equal length. In our winter the sun rises and sets in the south-east and south-west, in summer in north-east and north-west. The great fertility festival of the spring equinox was fixed for the day when the sun rises and sets midway between the two extreme positions which it occupies on the winter and summer solstices. Ancient monuments like Stonehenge and remains of the Maya calendar culture, show how the sun's position at sunrise or sunset on the solstices was fixed by alignment with two pillars of different height.

By the time when we first have written testimony to the cultural achievements of the ancient world, the priesthood had been long familiar with the precise direction of stars when they reached their highest altitude (transit) above the horizon. The south face of the Great Pyramid which was built about 2800 B.C. was so placed that the rays of Sirius at its transit struck it at right angles. One ventilating shaft leading to the royal chamber was so accurately placed that the dead Pharaoh was illuminated by the Dog star when it crossed the meridian. The main opening and a second shaft let in the light of the Pole star, Draconis, at its lower transit. The astonishing accuracy of these constructive feats was the fruit of centuries of recorded observations, and the means of recording the direction in which something lies still bear the impress of the physical reality from which the measurement of the angle took its origin.

Section I

(*a*) What particular aspect of measurement is stressed in the passage?
(*b*) What are the earliest signs of this type of measurement?
(*c*) What two fundamental lines of reference were settled before the great civilizations began?
(*d*) Which of these two lines was based on the noon position of the sun's shadow?
(*e*) Give two examples of the practical application by the ancients of the use of these lines of reference.

Section II

(*a*) Explain how the number of days in the year was fixed.
(*b*) Draw a diagram to show how the exact location of the noon shadow was fixed in ancient times.
(*c*) Demonstrate how the problem of the east-west meridian was solved.
(*d*) Explain the calculations involved in the construction of shafts in the Great Pyramid and illustrate this with a rough sketch.
(*e*) Explain the meaning and use in the passage of each of the following terms:

prerequisite	in "an essential prerequisite of settled city life"
heliacal	in "the heliacal rising of Sirius"
prime	in "the prime meridian"
orientation	in "orientation of burial sites"
vernal	in "the vernal and autumnal equinoxes"

Section III

Prepare the 'rough' of a wall-chart showing pictorially the information given in the passage. Pay particular attention to the chronological arrangement and to the way the main feature of measurement is emphasized.

7. In 1958 there was particular concern expressed about dangers to airline services, so crowded had they become. Radar systems were competitive and insufficient; standards of traffic control at some airports were too low. This extract from the journal *The Economist* describes the position. Read it for the facts it contains and the comparison made between British and other systems.

In many ways avoiding collision between two civil aircraft is the most serious problem that the controllers have to face. If aircraft are flying in civil air lanes, they will be flying much closer on each other's tails than if they were spread over the sky. They can converge on, and take off from, crowded airports like London and New York at the rate of one a minute; from there they may fan out across the North Atlantic where radio communication with them can easily be lost. Pressure of traffic demands more and more rapid handling of these aircraft, if they are not to waste time circling round the holding beacons while harassed controllers try to clear a path for them to land. And given the high fuel consumption of jet airliners at low altitude, they must be held, if they are held at all, at heights of 20,000 feet or more, from which they can swoop sharply down direct to the runway.

Radar has been the biggest boon to air-traffic controllers since the invention of radio, but there is not enough of it. Given a sufficient number of radar sets, a traffic controller can pinpoint each aircraft from the moment it taxis out of its hangar, through take-off, climb, and until it is fifty miles or so from the airport; then it can be picked up by a radar chain strung out along the air lane and monitored right through to its final landing. Such a radar chain can plot the position of every aircraft along the route, pick up any military jet that has strayed into the lane, and shepherd civil pilots whose navigation is going astray. Britain is the only country in the world where such a system is anywhere near realization. Pilots have for years given the Ministry of Transport unstinted praise for the efficiency of its ground control system, even though plans are far from complete. Until the Ministry has its own civil radar surveillance the whole length of the British air lanes five years from now, it is having to make do with help from the R.A.F.'s early warning radar.

The elaborate equipment used in Britain, such as the three

separate radar systems working at London airport, contrast oddly—and indeed alarmingly—with the limited tools that controllers are using in other countries. New York's splendid-looking Idlewild airport handles an aircraft a minute, but the controllers are doing a far more difficult job with much more rudimentary tools than in London. There, three civil airports and two military ones lie within a small circle. The big aircraft coming in from the North Atlantic have to thread a complicated pattern round and between these hazards with the help of one radar antenna that has to be re-tuned each time it is switched from long to short range. Yet New York is lavishly equipped compared with other airports.

Section I

(*a*) Write down the main topic of each of the three paragraphs.
(*b*) What has been the biggest boon to controllers since the invention of radio?
(*c*) Which Ministry is concerned with provision of control systems?
(*d*) What equipment is being used to help ease the situation?
(*e*) How many airports are concentrated in the New York area?

Section II

(*a*) Name two dangers to civil aircraft described in the first paragraph. Why do they exist?
(*b*) Enumerate the advantages of a radar chain.
(*c*) What is the state of radar development in Britain?
(*d*) No direct information is given about conditions at airports other than London and New York. What standards are *implied* by the writer?
(*e*) What do you consider to be the writer's purpose in presenting the facts about ground control conditions?

Section III

What is the present air control situation? List the changes that have occurred since 1958.

8. The following passage (an excerpt from an article by D. Peters on *Brakes* in the *Autocar*) deals with two types of car

braking systems—the self-servo and the leading-and-trailing shoe type. It is not an easy passage to read unless you are a motor-car enthusiast or a motor vehicle engineer, but you are required to gain from it an idea of the relative efficiency of the two types, and some idea of the advantages and disadvantages of each.

> Certain types of brake are more prone to fade than others, and these are the ones which are most efficient for their size. They are a class of brake which can be termed self-energizing and, as the term implies, they obtain quite a large amount of the effort required to apply them from the rotation of the drum itself. The design which makes most use of this is the self-servo brake, commonly found on American cars and familiar to British motorists in past years as the Bendix Duo-servo brake.
>
> In this type the shoes are moved by an expander, but the leading shoe is connected to the end of the second shoe instead of engaging with a fixed abutment. One of the shoes is moved towards the drum in the direction in which this is rotating, and the action of the drum is to drag the shoe with it. This drag is applied directly to the second shoe, and this in turn shares the self-applying effect. In this way total drag, and therefore braking effort produced, is great for a given size of brake, but the high shoe factor is the contributory cause of fade.
>
> By shoe factor is meant the ratio of drum drag to effort applied at the shoe by the wheel cylinder. If a large drag is produced for a relatively small effort—as in the self-servo type—then the ratio is large. A brake which relies less on drum rotation to help it to work requires a larger effort at the shoe to produce the required drag at the drum, and, therefore, has a low shoe factor. Hence a brake such as the leading-and-trailing-shoe type with a pair of shoes having a fixed anchorage and a common shoe expander, is an example of a low shoe factor brake, and this is commonly used for the rear wheels on British cars. Although more stable in performance than the self-servo brake, which sometimes has a tendency to fierceness and grab, it has the disadvantage that the leading shoe has a self-applying effort whereas the trailing shoe does not. The leading shoe, therefore, works harder, and the lining wears faster than that of the trailing shoe. There is also an unbalanced force on the drum because both shoes are not exerting the same outwards pressure.
>
> One solution, of course, is to make both shoes of leading type—that is to have a separate wheel cylinder for each and to move

them about a pivot in the same direction as drum rotation. A leading shoe produces a braking effort three times greater than that of a trailing shoe because of self-energizing action.

The problem of fade under extreme conditions remains, however, although the two leading shoe type is very efficient below fade temperatures. It is widely used for front brakes as its high energy absorption properties and compactness enable it to provide the required braking ratio between front and rear wheels, using the same size of drum and with light pedal pressure.

Section I

(*a*) Name the two types of brake discussed.
(*b*) Which of them is more prone to fade?
(*c*) Which of them has the higher shoe factor?
(*d*) Which type is used on the front wheels of British cars?
(*e*) What seems to be the main problem of both types of brake?

Section II

Interpret for the benefit of an amateur motorist with very limited technical knowledge, the following points. Set them out neatly, with sub-headings and diagrams where necessary:

The operation of a self-energizing brake
The advantages and disadvantages of a self-energizing brake
The operation of a leading-trailing-shoe brake
Its advantages and disadvantages
Comparison of the efficiency of the two types
Typical uses of each type of brake

Section III

Study the action and efficiency of a disc-type brake. Then present, graphically if you wish, a comparison between the disc brake and the two types mentioned in the article. Comment particularly on the relative efficiency and on the problem of fade.

9. This description of the preparation for sand casting is taken from Part I of W. A. J. Chapman's *Workshop Technology*. Read it straightforwardly for the information it gives, and concentrate on the order of operations. Consider, also, why the job of pattern-making is such a skilled one.

The Production of Castings

After bar and sheet, the next most common form in which material is received for working is in the form of castings. The casting of metal into the intricate shapes that are possible is an immense advantage in engineering construction, and the work of the foundry and pattern shop is of great importance. Castings may be made in steel, cast iron, brass, and bronze aluminium, as well as in numerous other alloys. A description of sand casting, which is the commonest method employed, will now be given.

A casting is produced by pouring molten metal into a mould made to the shape of the part required. In die casting this mould is made of metal, whilst in sand casting it is made in sand, its form being obtained by means of a wooden pattern having a shape similar to the part required. The first step, therefore, in making a casting, is that of making the pattern, and the pattern-maker may be counted as one of the most highly skilled craftsmen in the engineering industry. It is he who first has to interpret a drawing, however complicated, into the solid shape it represents. His experience decides upon the method to be used in the production of the casting, a factor upon which quality and success often depend. We will give a simple and obvious example. When metal is cast it contains gases which rise up through it, leaving at the highest point. The top surface of a casting therefore is liable to be rather spongy with blowholes left by entrapped gases unable to escape before the metal solidified. Now, if a round bar is cast on its side, the chances are that on one part of its circumference it may be a little spongy, whilst if it is cast vertically on its end, the sponginess will be at one end. If we extend this simple case to a complicated casting, we can see that the method of casting has some control over which portions of the casting will be sound and which may be spongy, and this method is under the control of the pattern maker by the manner in which he disposes the pattern. In our round bar, if the curved surface was important we should wish to have any likely blowholes in the end, and the pattern would be made for vertical casting. On the other hand, importance of the flat ends would make side casting preferable. Very often, castings can be so made that the spongy metal is in the portion of the casting which has to be machined, and it is removed by machining.

The wood mainly used for patterns is a well-seasoned soft wood such as yellow pine, and the pattern is made to conform with the outside shape of the casting. To allow for the contraction which takes place when the casting cools, the pattern must be a little larger than the finished casting. The amounts that different metals contract are as follows:

Metal	Contraction per foot
Cast iron	$\frac{1}{10}$ inch
Brass	$\frac{1}{5}$ inch
Steel	$\frac{1}{4}$ inch
Aluminium	$\frac{9}{32}$ inch

This contraction is allowed for by the patternmaker employing a special rule called a *contraction rule* with which to set off all the dimensions on the pattern. Since for cast iron a length of 12 in. contracts $\frac{1}{10}$ in. and cools to 11·9 in., a length of 12·1 in. will cool approximately to 12 in. On a contraction rule for cast iron, therefore, a length of 12·1 in. is marked as 12 in. and is then divided up exactly similar to an ordinary rule, so that when a patternmaker measures off a length with this rule, he automatically makes the length longer in the ratio $\frac{12 \cdot 1}{12}$ and so allows for contraction. For brass and other metals, the rule length is compensated in the same way according to their contraction.

The finished pattern will be smoothed up as much as possible to facilitate its withdrawal from the sand when moulding, and to impart a smooth surface to the casting. This effect is further increased by the shellac varnish with which the pattern is covered. When the casting is of a simple nature the appearance of the pattern is similar to that of the casting, but when the casting is more complicated with holes and cavities it becomes necessary to resort to coring in the mould. This will be best explained when we are discussing the actual moulding process.

Section I

(*a*) What is the difference between die and sand casting?
(*b*) What causes the metal in a casting to become spongy?
(*c*) What kind of wood is usually chosen for patterns?
(*d*) What is the contraction per foot of cast iron?
(*e*) Why is the pattern given a smooth finish?

Section II

(*a*) Explain why the patternmaker needs such high skill.
(*b*) How are spongy surfaces arranged or avoided?

(*c*) What is a contraction rule? Draw a rough plan of a contraction rule for steel castings.

(*d*) Set out briefly the main steps in the preparation of a pattern for sand casting.

(*e*) Draw a series of diagrams illustrating the example given by the writer of sponginess in the casting of a round bar.

Section III

Refer to the actual casting process and write a short description, with diagrams, of the overall process from the pattern-making to the cooling.

10. This extract is from *Basic Electronics* by R. A. Greenfield. It is an introduction to a chapter on Semiconductors and gives a very brief idea of the background to this new development. Note particularly the discovery and rejection of the crystal in the early days of electronics.

Within the past few years, semiconductor devices have opened up an entirely new field in electronic engineering. Transistors are now used in computers, control, and servo-systems, amplifiers, and oscillators, as well as in circuits employing special semiconductor devices, relay units, photoconductive cells, and phase-sensitive detectors, to mention just a few. The thermionic valve, hitherto one of the principal circuit elements employed in radio engineering to control the flow of electricity, has now been supplemented by semiconductor devices. . . .

The history of semiconductors begins in 1906 when Pickard patented the crystal detector—a cat's whisker of steel wire in contact with a crystal of galena. The galena crystal behaved as a rectifier of radio-frequency signals. About this period much empirical data was available on the rectifying action of crystals such as zincite, carborundum, pyrites, and other crystals. Little knowledge was available on the nature and structure of solids. These crystals were a practical realization of the fact that some solids exhibited the property of asymmetric conduction of electricity, a phenomenon first observed in 1835. The thermionic valve discovered in 1903 was a better proposition to the electronic engineer because it possessed the property of amplification as well as that of detection. The crystal lacked the property of amplification and was accordingly considered destined for the museum shelf.

For the next thirty years, much research was directed towards a better understanding of the structure of the solid state, especially the relationship of electricity and matter. During this period, the fundamental theories on what is termed today 'the physics of the solid state' were founded.

Towards the end of this period, the electronic engineer developing ultra-high-frequency equipment found that the thermionic valve had limitations. Crystal diodes of silicon and of germanium were particularly suitable for the detection of these signals. The crystal device still lacked the property of amplification. In 1941, the P-N junction was discovered, and in 1948, the first point-contact transistor, a semiconductor device capable of amplification invented by Bardeen and Brattain, appeared on the market. This was followed by the discovery of the P-N-P junction transistor by Shockley in 1949; the phototransistor, using a point-contact on N-type germanium, in 1950; and, in 1951, Shockley again reported the discovery of the N-P-N transistor. The past few years have seen many improvements in design, and other elements are still being examined with a view to their possible use in semiconductor devices.

Section I

(*a*) Give two examples of the use of transistors.
(*b*) What type of valve has the transistor begun to replace?
(*c*) What was the main limitation of the early crystal?
(*d*) At what date does the story of the semiconductor begin?
(*e*) What equipment led to the demand for semiconductors?

Section II

(*a*) Although the crystal detector was patented at the beginning of the century, why was it not developed and why did electronic engineers prefer the thermionic valve?
(*b*) How do you account for the break of thirty years before the re-discovery of semiconductors? Did anything of value develop during that time?
(*c*) When was the problem of amplification overcome and by whom?
(*d*) What is the purpose in the circuit of either a thermionic valve or a semiconductor?
(*e*) Explain the meaning and use in the passage of the following words or phrases:

rectifier	in "a rectifier of radio signals"
empirical	in "much empirical data was available"
asymmetric conduction	in "the property of asymmetric conduction of electricity"
ultra-high-frequency	in "ultra-high-frequency equipment"
point-contact	in "point-contact transistor"

Section III

(*a*) What is the rest of the story behind the phenomenal change caused by the transistor? List the advantages it has over the ordinary valve and give some idea of the widespread effect that it is having on industry and society.

(*b*) Prepare a draft chart showing the two stages of crystal development, contrasting the old cat's whisker with the new transistor radio. Make your diagrams vivid and interesting.

VI

NOTE-TAKING

Note-taking should be a process of investigation and analysis. The trained research worker keeps his notebook by him and records all relevant information methodically and according to the purpose of his investigation. For the student engineer note-taking is too often an aimless task, a picking up of bits of information in magpie fashion, or the routine copying of notes from the blackboard or at a lecturer's dictation. These methods are calculated to stifle valuable training.

Note-taking is an important training in judgement and in the art of selecting important facts and ideas from a welter of general information—the investigation of a process, the extracting of relevant information from a lecture or paper, the detailed results of testing and the conclusions drawn from those results.

Notes are consequently made within a framework. They are virtually the re-creation of the main ideas in the writer's or lecturer's mind, and these main ideas must be appreciated by the student whatever details he may choose to retain within the framework.

A set of notes is not just a short version of the original. Take these two extreme examples:

(*a*) *Brief notes.* A man set to watch a dial may be asked to note the reading every three hours. Although the process may take twenty-four hours, the notes will consist merely of a set of figures.

(*b*) *Extensive notes.* As part of a psychological experiment a man may be locked in a blackened room and asked to make comments upon his own feelings and reactions. These may be dictated into a tape recorder, and the results, when typed, may run to many pages.

It is important to have the right attitude to notes. Remember that they are not aimless jottings; you must

(*a*) have a definite purpose
(*b*) be aware of the limits of that purpose
(*c*) keep the overall subject in mind as you select facts
(*d*) keep your notes in neat order, using headings, subheadings and enumeration
(*e*) be prepared to amend your division suddenly if the lecture or book is badly organized.

MAKING NOTES FROM YOUR READING

Do not plunge into the material making haphazard notes of what *seems* to be important. You will have only a shapeless mass of notes which will be less readable than the original. Be warned by the keen note-taker who started reading a book with the intention of underlining the important points. Because he was keen but had no real plan he finished up with most of the sentences in the book underlined, and with himself little the wiser for his reading.

Make your notes in the following way:

(*a*) *Skim the passage or book (use the list of contents) to discover:*

the theme
the scope
the line of development

(*b*) *Note down the main divisions and construct suitable headings. Use existing headings if they help.*
(*c*) *Under each heading make brief but clear notes of the main points, or those points which you wish to retain.*

REMEMBER:

Know what you want from the passage. You may even be able to find it without reading in detail.
Do not be put off by irrelevant material, however interesting.
If something catches your eye make a brief reference note for further reading.
Careful preparation before writing notes means that you can avoid a rough note stage.

Read the following extract from W. Steeds' *Engineering Material, Machine Tools and Processes*. It deals with the classification of machine tools and is the sort of material that you might need to reduce to note form for easy memorizing before an examination. When you have read the passage, study the example of note-making that follows it.

Machine tools are now so numerous in type that it is impossible to describe every separate type in a book of reasonable size; nor is this necessary, since the differences between many of them are merely matters of detail and not of principle. It is possible to classify machine tools into a comparatively small number of basic groups and it will be sufficient to describe, fairly fully, an example of each group and to consider the chief variations within the group in rather less detail. The classification can, however, be done in several ways, although none of them gives absolutely clear-cut groupings.

One possible classification is into four groups—*single-purpose*, *multi-purpose*, *special*, and *transfer*. A single-purpose machine is one that is designed to do one particular kind of operation but which is built more or less as a standard product by a number of machine-tool builders. The single-purpose machine is not, however, restricted to one particular job; for example, a shell-turning lathe is designed specially for shell turning and is not adapted for general lathe work; it is not, however, restricted to one particular size or profile of shell. Single-purpose machines are generally simple in design and are often little more than standard multi-purpose machines stripped of all the 'frills' and simplified as much as possible; they are consequently low in first cost. Unless they can be occupied for a reasonable percentage of their life on the job for which they are designed, it would generally be better to use a multi-purpose machine which can do other jobs as well.

A multi-purpose machine tool is one that is capable of doing a number of different types of operation. For example, a sliding, surfacing, and screw-cutting lathe fitted with a taper attachment can do almost every lathe operation except such specialized ones as the relieving of milling cutters and hobs and the turning of non-circular sections. The multi-purpose machine costs more than the single-purpose one, but in most machine shops it can more easily be kept fully occupied and it is consequently the most widely used type. When the number of different operations possible is large, the multi-purpose machine may sometimes be referred to as *universal*.

The special machine tool is one built specially for some particular job that is required in large quantities. It has to be specially designed for that job and is not usually capable of doing any other job, so that it must be scrapped when the job ceases. It costs more than the single- or multi-purpose machines but does the job in less time and this is able to show a saving in total cost if the number of articles required is sufficiently high.

The transfer machine is, in effect, several machines coupled together, and operated automatically as a single unit. They are of various types, the principal ones being described later. . . . The first cost of a transfer machine is usually high, although it may sometimes be less than the total cost of the several machines it will replace, but the operating costs are low because fewer operators are required. They are 'specials' in that they are built to order and for a specific series of operations on a single component and are essentially high-production equipment.

A second method of classification is according to the kind of cutting tool used, though it would be more precise to say according to the size of chip removed. This classification gives two groups, namely, machines using 'cutting tools' which produce chips visible as such to the unassisted eye and machines using abrasives which produce 'chips' so small that they can be recognized only with the aid of a microscope. Until a few years ago the first group could have been defined as machines using steel cutting tools, but cutting tools are now frequently not made of steel.

A third method of classification is according to the type of surface principally produced—*e.g.*, surfaces of revolution, plane, or ruled surfaces and miscellaneous surfaces. Thus lathes are primarily adapted for producing surfaces of revolution and milling machines for producing ruled surfaces. This method is the least satisfactory of the three methods considered but has some advantages.

There is no space in the above classifications for spark, supersonic, chemical or electrical erosion machining methods but all of these have been tried experimentally and the first two are now coming into general use.

Now we shall go through the process of making notes:

1. *Skimming*

A quick reading of the passage will have shown that the writer deals with three ways of classifying machine tools, and that he gives some explanation of the types of machine in each group. This gives a paragraph plan as follows:

Para. 1 Introduction
2 Four-group classification: (*a*) single-purpose
3 (*b*) multi-purpose
4 (*c*) special
5 (*d*) transfer
6 Cutting-tool classification
7 Surface classification
8 New methods not classified.

The main emphasis is on the four-group classification.

2. *Main Headings*

These should show the classifications—three, together with a mention of unclassified machines. This suggests four main sections which can be indicated by Roman capitals I, II, III, and IV:

E.g. I *Four-group classification (by function)*
II *Classification according to size of chip*
III *Classification by surface*
IV *Non-classified machines*

3. *Sub-headings*

Section I falls naturally into four parts. We can use ordinary numerals for these:

I *Four-group classification (by function)*
1. Single-purpose
2. Multi-purpose
3. Special
4. Transfer

Sections II and III will be similarly divided, and a division can be made of Section IV into those coming into general use and others still experimental:

II *Classification according to size of chip*
1. Cutting tools
2. Abrasives

III *Classification by surface*
1. Surfaces of revolution
2. Plane surface
3. Miscellaneous surfaces

IV *Non-classified machines*
1. Spark and supersonic
2. Chemical or electrical erosion

4. *More detailed Notes*

It may well be that the above list of headings and sub-headings will be sufficient for your needs. If not the notes can be further expanded as follows. This time we add a title

Classification of Machine Tools

I *Four-group classification (by function)*

1. *Single purpose:* One particular operation
   Low first cost
2. *Multi-purpose:* Number of different operations
   High first cost; but can be kept fully occupied (sometimes called *universal*)
3. *Special:* Built specially for a job requiring large quantities. Scrapped later
   Higher first cost, but saving in mass production
4. *Transfer:* Several machines coupled. Automatic as one unit
   High first cost, but low operating cost with fewer operators
   High-production equipment

II *Classification according to size of chip*
1. *Cutting tools:* visible chip
2. *Abrasives:* chip microscopic

III *Classification by surface*
1. *Surfaces of revolution: e.g.*, produced by lathe
2. *Plane surfaces: e.g.*, produced by milling process
3. *Miscellaneous surfaces*

IV *Non-classified*
1. *Spark and supersonic:* Coming into general use
2. *Chemical or electrical erosion:* Still experimental

N.B. This is only one example of the way in which notes can be made of this particular passage. You will notice that I have not included personal comments made upon the classifications by the writer, neither have I included some of the specific examples of machine types quoted in the original. There is an argument, also, for arranging the notes in another framework, such as:

I Main Classification by Function
II Other Classifications
III Non-classified

Now apply the methods of note-taking in the following exercises. Remember always to skim through for the *framework* of the passage first.

Exercises in Note-taking

Each of these exercises will be prefaced by instructions and advice, in an attempt to give some sense of purpose to the work. In the earlier exercises you should make a rough pencilled draft of your main and sub-headings before going on to the main notes. It is important that you display the notes neatly, using underlining and enumeration where useful. Use diagrams if they are helpful.

1. The following passage deals with the structure and the process of producing malleable cast iron. It may be familiar to you already. Make a set of brief notes that would act as an *aide memoire* for answering an examination question on the subject.

Malleable Cast Iron

A malleable metal is one which may easily be caused to spread and flatten under pressure or hammering. Lead is the best example of a malleable metal we have in the workshop, and the reader may soon prove for himself that this metal is easily beaten out and flattened.

The application of the term *malleable* to castings of that type is rather a misnomer because they are not very malleable when compared with the usual standards of malleability. When compared with grey iron castings, however, which are fairly brittle, malleable castings do possess a degree of toughness, and this is probably why they have been so named.

For many purposes it is necessary to produce small, thin parts having a reasonable degree of strength, such as would not be obtained in a grey iron casting which, as we have seen, is brittle and would break under rough usage. Such parts could be built up or forged from steel, but it is much cheaper to produce castings, and the process of converting castings to the tougher form of malleable iron has been evolved to meet the need for such products.

We have seen that the structure of cast iron contains flakes of graphite dispersed through it, and it is the breaking up of its continuity by these rather large flakes of weak material that makes the iron brittle. If the carbon were all removed there would be left wrought iron, a ductile material which bends easily. Obviously, we could not subject finished castings to a process similar to that used for the manufacture of wrought iron, but if we could use some of the carbon and cause the remainder to be broken up from its flake-like form into tiny specks it would not have such a weakening effect, and our casting should not break when dropped. This, in general, is what takes place in the process of making malleable castings. First, castings are made from an iron having most of its carbon in the combined form. Two methods are then used for malleabilizing the castings: the Whiteheart, used in England, and the Blackheart, more common in America. The names refer to the colour of the fracture given by castings produced by each method.

In the Whiteheart process the castings, composed of cast iron with most of its carbon in the combined state, are packed in iron or steel boxes and surrounded with a mixture of used and new haematite ore. The boxes are heated to a temperature of 900–950° C. and maintained at that temperature for several days, during which time part of the carbon is oxidized out of the castings, and the remainder is dispersed in small specks throughout the structure. In thin sections the structure might be almost pure iron, but in thicker sections the outside approaches to nearly pure iron and the centre has the dispersed specks of carbon. The heating period is followed by a very slow cooling which occupies several more days and the result is a casting which is tough and which will stand hard treatment without fracture. The tensile strength of malleable castings is about 25 tons per square inch.

Workshop Technology, Part I, by W. A. J. Chapman

2. A significant event in the history of lighting was the introduction of the arc lamp. Its effect upon agriculture and industry in the nineteenth century is shown in the following extract from William O'Dea's *The Social History of Lighting*. When you have read the extract, make outline notes under section headings. You are advised to construct your framework carefully before filling in the notes.

It was the advent of electricity that made the most revolutionary changes in the lighting of work after dark. The arc lamp was demonstrated in a laboratory in 1810, but there were no machines capable of generating the electricity required. Surprisingly enough this was not sufficient to preclude entirely the use of the arc lamp as an aid to work by night. At great expense and difficulty there were one or two attempts to feed the arcs with great batteries of Bunsen cells. They were quite uneconomic. A quarry at Angers was arc-lit in 1863, very soon after primitive machines became available. The Guadarrama section of the Spanish Northern Railway was cut about the same time and it was estimated that arcs were in use for 10,000 hours in the process. It was said that the Paris Exhibition of 1878 could not have been opened on time had arc lamps not been used by the builders, but the efficiency and reliability of machines for generation, such as the Gramme and Siemens types, was by then greatly improved. A couple of years later M. Albaret of Liancourt devised a system of working on the harvest after dark by using lamps on high supports each of which was reputed to light effectively an area of 100 metres radius. The portable steam engines driving the dynamos could also be used as a source of power for the threshing machines. These agricultural applications were not widely employed, but for building work the arc lamp sprang into immediate favour and only in recent times have improved high-power light sources supplanted the arc.

For indoor applications the arc lamp was clearly unsuited to ordinary rooms, but for large, high-ceilinged areas it had much to commend it, in view of the defects of contemporary lighting methods. In France, M. Menier had his chocolate factories lit electrically in 1875, and both the Compagnie Chemin de Fer du Nord in 1876 and the P.L.M. Railway in 1879 lit their main parcels halls. In London and Glasgow, King's Cross and St. Enoch's stations followed suit shortly afterwards. Each of these installations involved the setting up of a generating station as there were no public supplies. The lighthouse projector works of Sautter & Lemonnier, another example of a high-roofed area, was among the pioneer factory installations in France.

The railway companies were so early in the arc-lighting field that it may be thought surprising there were not arcs on high towers to light up marshalling yards. The reason was probably one of economics. The attraction of brightly lit main-line stations serving the travelling public would be worth considerable expense, both for installation and subsequent running costs, but gas still remained cheaper; also great numbers of low-set gas lamps were already installed at the shunting yards and it would have been a costly matter to scrap them. Later it proved economical to use the same installations converted to electric filament lamp lighting and although supplementary arc lamps were also employed they were not without disadvantages. High mounting heights could be a great disadvantage in murky weather and the trend even today is towards large numbers of moderately powered mercury or sodium lamps at medium heights.

For factory work the arc also had many disadvantages. Even with the best of regulators its constancy left much to be desired and it threw the harshest of shadows. It had to be suspended well away from the work and was usually troublesome to get at when it required attention. The less efficient incandescent electric lamp might have sounded a better proposition were it not that the more cheaply run arc lamp nevertheless cost much more than gas lighting. Swan ran his own lamp factory in Newcastle by the light of Swan lamps in 1880–1 but as his 20-candlepower lamps, in their holders, cost 25s. each at first, and lasted only a very few hundred hours, the project was hardly economical. By 1883 the price had come down to 5s. a lamp, but even that made them suited more to lighting rich men's houses than factories. . . .

It was many years before incandescent filament lamps ousted the arc from more general favour. About 1885–6 the Brush Company, in the United States, ran many power stations supplying arc lamps. It was reckoned that a demand of more than 100 lights became economical and the tariff was £30 ($150) per lamp per year for factories, yards and similar private uses. The arcs were only switched on from 5-10 p.m. in winter, 7-10 p.m. in summer and 1-2 hours on Saturday. For a lamp burning all day 4s. per day was charged, but this fell to 2s. 9d. per lamp if there were as many as 25 in circuit. The consumer was not charged for connection to the supply and night watchmen controlled the switching on and off for municipal authorities who were charged only 2s. 6d. to 3s. per lamp per night for street lighting.

3. The following passage, taken from the Oxford *History of Technology*, gives a brief history of the bicycle. Compile a set of notes, clearly arranged in chronological order, showing the development of the machine.

The basic idea of a two-wheeled machine astride which a man might sit and propel himself along is undoubtedly of considerable antiquity, but it did not assume a practical form until 1818. Then a German, Baron Karl von Drais de Sauerbon (d. 1851), constructed an improved type of hobby-horse, or dandy-horse, as such machines were called in England. It consisted of a wooden frame or backbone, to which two wheels of equal size were attached by iron brackets, the front one being capable of turning relatively to the frame by means of a handle-bar fixed to the top of the front fork. The rider sat on a saddle mounted in the centre of the backbone and leaned forward against a padded support between the saddle and the handle-bar, propelling himself by taking long strides with each foot in turn against the ground. The wheels were wooden and no brake was fitted. To ride the hobby-horse downhill must have been an alarming experience.

About 1839 Kirkpatrick Macmillan, a Scot, built a machine incorporating a system of treadles, linked by connecting rods and cranks to the rear wheel, which was the first bicycle to be propelled by the rider without touching the ground with his feet. A brake on the rear wheel could be applied by twisting the handle-bar. It was not until 1861 that cranks and pedals were directly applied to the wheels of a hobby horse, and credit for the first use of these on the front wheel must go to Ernest Michaux, a Frenchman, who by 1867 had formed a company for the manufacture of velocipedes of his own design. Late in the previous year the first American bicycle patent had been taken out jointly by Pierre Lallement and J. Carrol.

During the next few years the bicycle industry in France grew rapidly. On November 17, 1869 a great international race from Paris to Rouen, a distance of 83 miles, was won by an Englishman, James Moore, riding a machine fitted with ball-bearings. He subsequently designed more efficient ball-bearings, which were soon adopted by manufacturers generally. By this time England was waking up to the fact that there was a growing demand for bicycles, and in 1869 the first machines to be made in this country were built at Coventry. James Starley (1831–81), of that city, took out a patent for the first ladies' bicycle in 1874, and also invented the tangentially spoked wheel—a great improvement on the earlier radial-spoke design. He patented the *Coventry* tricycle in 1876 and a year later took out another patent for the use of the differential gear in combination with chain-drive, which features he embodied in the *Coventry* tricycle. This machine paved the way for the introduction by other manufacturers of many different designs with two, three and four wheels. More than 200 varieties were on the market in 1884.

The high or 'ordinary' bicycle was now rapidly becoming the standard type, for a reason which is comprehensible if not immediately obvious. In order that each revolution of the pedals attached directly to the hub of the front or 'high' wheel should cause the machine to travel a reasonable distance, it was necessary that the wheel should be from 4 to 5 ft. in diameter, and, in fact they were so made, with variations in steps of 2 in. to suit the length of the legs of different riders. Attempts to make the front-driving machine safer were numerous, one such being the *Kangaroo* bicycle in which the front or driving wheel was considerably smaller than in the case of the 'ordinary,' the pedals being mounted at a point about half-way between the hub and the ground and driving the former by geared-up chains. Another design was the *Geared Facile* which incorporated a lever action and sun-and-planet gearing. A slightly inclined front fork also helped by placing the weight of the rider farther behind the centre of the wheel.

The advent of the contemporary safety bicycle was foreshadowed as early as 1876, when H. J. Lawson patented a rear-driving machine known as the *Crocodile*. In this machine the rear wheel, rotated by a system of levers and treadles, was considerably larger than the front. It was not a commercial success. In 1879 Lawson designed the first rear-driven safety bicycle using ordinary cranks and pedals, a geared-up chain-drive to the small rear wheel being employed. A large front wheel gave the machine a rather ungainly appearance.

It was left to J. K. Starley of Coventry, nephew of James Starley, to design and manufacture the first commercially successful safety bicycle, the *Rover*, in 1885. It employed all the elements of the modern machine, having wheels of almost equal size, direct steering with inclined forks, and geared-up chain-drive to the rear wheel. It soon attained great popularity. The introduction of this machine and others constructed on similar principles doomed the 'ordinary' to extinction. The invention of the pneumatic tyre very soon afterwards was the final blow, and by the turn of the century the safety bicycle, now fitted with a free-wheel, a device that had been anticipated at least thirty years earlier, reigned supreme.

4. Samuel Smiles, a nineteenth-century biographer, wrote in praise of Henry Maudslay the metal worker who did so much to improve, among other things, the operation of the lathe. In the following extract a picture is given of the situation before the development of the slide rest, and the effect that its introduction had upon engineering. Remember that Smiles was writing at a time when the results of Maudslay's labours

were still fresh. Make brief notes to show the story of the development of the slide rest and its importance to modern industry.

It is unquestionable that at the time when Maudslay began the improvement of machine-tools, the methods of working in wood and metals were exceedingly imperfect. Mr William Fairbairn has stated that when he first became acquainted with mechanical engineering, about sixty years ago, there were no self-acting tools; everything was executed by hand. There were neither planing, slotting, nor shaping machines; and the whole stock of an engineering or machine establishment might be summed up in a few ill-constructed lathes, and a few drills and boring machines of rude construction. Our mechanics were equally backward in contrivances for working in wood. Thus, when Sir Samuel Bentham made a tour through the manufacturing districts of England in 1791, he was surprised to find how little had been done to substitute the invariable accuracy of machinery for the uncertain dexterity of the human hand. Steam-power was as yet only employed in driving spinning-machines, rolling metals, pumping water, and such like purposes. In the working of wood no machinery had been introduced beyond the common turning-lathe and some saws, and a few boring tools used in making blocks for the navy. Even saws worked by inanimate force for slitting timber, though in extensive use in foreign countries, were nowhere to be found in Great Britain. As everything depended on the dexterity of hand and correctness of eye of the workmen, the work turned out was of very unequal merit, besides being exceedingly costly. Even in the construction of comparatively simple machines, the expense was so great as to present a formidable obstacle to their introduction and extensive use; and but for the invention of machine making tools, the use of the steam engine in the various forms in which it is now applied for the production of power could never have become general.

In turning a piece of work on the old fashioned lathe, the workman applied and guided his tool by means of muscular strength. The work was made to revolve, and the turner, holding the cutting tool firmly upon the long, straight, guiding edge of the rest, along which he carried it, and pressing its point firmly against the article to be turned, was thus able to reduce its surface to the required size and shape. Some dexterous turners were able, with practice and carefulness, to execute very clever pieces of work by this simple means. But when the article to be turned was of considerable size, and especially when it was of metal, the expenditure of muscular strength was so great that the workman soon became exhausted. The slightest variation in the pressure of the tool led to an irregularity of surface; and with the utmost care on the workman's part, he could not avoid occasionally cutting a little too deep, in consequence of which he must necessarily go over the surface again, to reduce the whole to the level of that accidentally cut too deep; and thus possibly the job would be altogether spoiled by the diameter of the article under operation being made too small for its intended purpose.

The introduction of the slide rest furnished a complete remedy for this source of imperfection. The principle of the invention consists in constructing and fitting the rest so that, instead of being screwed down to one place, and the tool in the hands of the workman travelling over it, the rest shall itself hold the cutting tool firmly fixed to it, and slide along the surface of the bench in a direction exactly parallel with the axis of the work. Before its invention various methods had been tried with the object of enabling the work to be turned true independent of the dexterity of the workman. Thus, a square steel cutter used to be firmly fixed in a bed, along which it was wedged from point to point of the work, and tolerable accuracy was in this way secured. But the slide rest was much more easily managed, and the result was much more satisfactory. All that the workman had to do, after the tool was firmly fitted into the rest, was merely to turn a screw handle, and thus advance the cutter along the face of the work as required, with an expenditure of strength so slight as scarcely to be appreciable. . . .

The effects of the introduction of the slide rest were very shortly felt in all departments of mechanism. Though it had to encounter some of the ridicule with which new methods of working are usually received, and for a time was spoken of in derision as, "Maudslay's Go-cart"—its practical advantages were so decided that it gradually made its way, and became an established tool in all the best mechanical workshops. It was found alike capable of executing the most delicate and the most ponderous pieces of machinery; and as slide lathes could be manufactured to any extent, machinery, steam engines, and all kinds of metal work could now be turned out in a quantity and at a price that, but for its use, could never have been practicable.

In course of time various modifications of the machine were introduced—such as the planing machine, the wheel-cutting machine, and other beautiful tools on the slide-rest principle, the result of which has been that extraordinary development of mechanical production which is so characteristic a feature of the age we live in.

Industrial Biography, by Samuel Smiles

5. The rôle of plastics in modern industry is an important one. The following extract—from Arnold Allcott's *Plastics To-day*— shows the wide-ranging usefulness of these products and their relationship with the traditional construction materials. Under the heading *The Uses of Plastics*, tabulate the information given in the passage. Use sub-headings and column arrangement if necessary.

Many of the traditional constructional materials, which have long served and will continue to serve so many useful purposes, suffer from certain disadvantages. Iron and steel are liable to rust, and most metals undergo corrosion, either on exposure to industrial fumes and gases present in the atmosphere or when in contact with liquids containing acids, alkalis, or other active substances.

Wood may be damaged by fungus, dry rot, or woodworm, and in exposed positions requires periodic painting for protection.

In plastics we have a class of material with collective properties not possessed by anything found in nature.

They do not rust or decay, are substantially immune from attack by insects or vermin, and are in general not affected by moisture, oils, acids, and alkalis. The majority are not much heavier than hardwood, and some types are, weight for weight, as strong as most metals. In addition, all plastics are practically non-conductors of heat and electricity. Some kinds can be produced in filament form and woven into cloth which is moth-proof and creaseproof.

It is not surprising that such versatile materials have not only proved to be satisfactory alternatives to many traditional materials, but have in some cases been able to serve purposes for which no previously existing substance was suitable. This does not mean that plastics are likely to replace metals, wood, bricks, or stone as major constructional materials, in spite of the sensational prophecies which have been made from time to time by enthusiastic but ill-informed writers. Outstanding as their properties are they should not be regarded as universal substitutes, and attempts to use them as such and with conventional means of fabrication have frequently resulted in grave disappointment.

The true rôle of plastics in industrial application is that of supplements to the older materials, and although in some cases, notably in the electrical industry, plastics have partly superseded the older materials it is rather as components than as complete substitues that their chief value lies.

The electrical engineer uses plastics for fuse boxes, switch, and plug bodies, and for cable insulation and protection. The mechanical engineer can now use silent-running gear wheels and heavy duty bearings made in plastics. These are hard wearing and will run undamaged in water, oil, or petrol. The motor industry is making an increasing use of plastics for moving parts, brake linings, interior fitments, upholstery, and in some cases for car bodies.

Conduits for corrosive liquids, protective clothing, food packs, and general packaging are some of the almost universal uses of plastics. Even in the foundry plastics play their part, being used as core binders and also in a modern development known as shell moulding, in which a mixture of dry resin and sand is poured over a hot metal pattern to form a mould which is then hardened by further heating in a stove. This gives a strong mould which is many times lighter than the traditional mould, can be made in a matter of minutes, and enables the moulder to work to closer tolerances.

The plastics industry now provides strong adhesives which, unlike animal glue, are not affected by moisture or fungus. Resin bonded plywood can be used for outdoor purposes or even under water. The fuselage of the famous Mosquito aeroplane which was used for bomber and fighter sorties during the Second World War was made of layers of wood bonded together by a plastics adhesive, a striking example of the happy association of old and new constructional materials. Similar adhesives are used for metal-to-metal bonding, and in many cases it is considered that this method is superior to riveting, as the stresses are spread over the whole bonded surface instead of being concentrated round the rivets, so that the failure load may be as much as 40 per cent. greater than that with a riveted bond.

The general public is also well catered for. The housewife has a choice of domestic utensils such as the baby's bath, the washing-up bowl, jugs and measuring vessels, cups, saucers, plates, and tumblers, in fact all kinds of table crockery. These articles are light and not easily broken, easy to keep clean, can be produced in many pleasing colours, and there is no danger that they will scratch or chip porcelain enamelled sinks. It is true that they must not be used in hot ovens or over the flame of a gas stove, but they are not harmed by hot water.

Plastic draining boards, table tops, floor coverings, and wall tiles are useful and pleasing adjuncts to the modern kitchen. In the garden a plastic hosepipe will probably last longer and stand more ill treatment than a rubber one. Transparent plastics may be used instead of glass for frames and cloches, and light plastic sheets for soil mulching or plant protection.

In the past there have been cases of shoddy or unsuitable articles coming on the market. This was perhaps inevitable in a new and exciting industry, which no doubt attracted some operators who lacked appreciation of the true potentialities of the new materials and who had little or no experience of the requisite methods of fabrication.

Fortunately this unhappy period is past. Reputable and experienced manufacturers are able to put their products to stringent tests both during and after fabrication. A series of standards specifications has been evolved both for moulding materials and finished products. . . .

There are so many ways in which plastics can be of great benefit to mankind that it is unwise to risk disappointment by asking them to perform functions for which they are not suitable.

TAKING NOTES FROM LECTURES

Do not depend upon cyclostyled 'hand outs.' Develop the ability to follow and analyse a lecture and to note down the main points. Not only will this keep you awake, it will also help you to develop a critical attitude.

The success of your note-taking is, naturally, affected by the coherency of the lecture. A rambling and discursive lecture is always difficult, and sometimes impossible to reduce to note form. A typical well-organized lecture will probably have some such framework as this:

Introduction of subject
Explanation
Repetition of main points to ensure it has gone home
Summing up and practical application.

This is only one of many frameworks, and you should try to identify the particular method used by the lecturer. This will help you to secure your main heading early in the lecture.

If the lecturer introduces his subject plainly and tells you how he intends to develop it, you would be wise to put down a series of rough headings (with the hope, of course, that the lecturer will actually keep to them).

It is not possible to provide direct exercises in this book, but here are a few do's and don'ts which may help:

DO NOT
(*a*) panic. You will never make a coherent note if you do.
(*b*) take down every interesting little fact as it comes along. You may eventually be writing everything that is said.
(*c*) try to write down everything verbatim.
(*d*) make notes so intensely that you find that you lose the thread of the lecture.
(*e*) be put off by the lecturer wandering from the point. Learn to recognize this, and put your pencil down until he gets back.
(*f*) try to make notes with only one pencil or pen.
(*g*) use abbreviated forms which you may not understand later.

DO
(*a*) be prepared by giving some thought to the topic beforehand.
(*b*) listen carefully for the key introductory phrases.

e.g. "I am going to deal with *three properties* . . ."
"There are two ways of . . ."
This will provide your sub-headings, etc.

(*c*) space your notes reasonably. Use a proper notebook, not the backs of envelopes.
(*d*) number your notes, especially sequences and lists.
(*e*) keep awake.

EXERCISES

1. Arrange for certain students in your group to prepare short lectures which they then deliver while the rest of you make

notes. Compare your notes with the plans made for the lectures.

2. Make use of B.B.C. programmes. Select a talk in which you are interested and try to make notes as you listen. If the talk is printed later in the *Listener* you will have a chance to compare your notes with the original.

VII

SUMMARIZING

Whatever the name you may give to the process—abstracting, précis-writing, summarizing, reducing, digesting—it is a useful technique in a world that depends increasingly upon a heavy load of communication. Reports, trade magazines, data sheets, journals, bulletins, minutes—all these present a formidable reading problem to the scientist or technologist. Sometimes the material can be rendered in note form; sometimes the material needs to be reproduced in a shortened but readable version specially designed for another reader.

Summarizing not only saves words, it saves time and money. An executive may not have time to wade through an extensive research report before making a policy decision. A summary of the main findings and recommendations will leave him in a happier and more decisive mood. Abstracts from certain technical articles may enable a busy engineer to keep up to date with modern methods. Already specialist staffs deal with this re-presentation of information in shortened form, selecting from the general mass of material those items which are of interest and concern to the particular reader.

The process of summarizing is not an easy one and there is no quick and easy way to develop the technique. It involves

(*a*) the ability to read efficiently
(*b*) the ability to understand
(*c*) the skills of judgement and selection
(*d*) the art of interpretation and re-presentation

You will have practised some of these techniques already. Much of the work in the sections dealing with comprehension and interpretation and the art of note-taking has been a preparation for the exercises to follow.

Method

Most of you will already have received some instruction in formal précis work. It may therefore be as well for you to base your work on the method you have been taught: but please do not let *method* interfere with the purpose of your summary and the attention that should be given to the intended reader.

Remember that a summary should be:

(*a*) a selection of the main facts or ideas, or the facts required by the reader.
(*b*) your re-presentation of these ideas (*not* a line-by-line pruning and re-hashing of the original).
(*c*) a connected, grammatical piece of writing.

EXERCISES

Each of the exercises will be preceded by a brief introduction which will give an idea of the content of the extract and the purpose of summarizing.

1. The following short extract from Arthur Clarke's *Interplanetary Flight* describes the possible design of "spacesuits." You are required to convert the description into a brief specification of the two types of suit; this should be suitable for entry in a 'space dictionary.'

> For the exploration of airless bodies like the Moon, or for such operations as orbital refuelling or assembling structures in space, it would be necessary to design what have been christened 'spacesuits' so that men can exist and work under considerable extremes of temperature and in a vacuum. Such suits would have much in common with a modern self-contained diving dress but would have to meet more stringent requirements. They would carry enough oxygen for several hours, and would be fitted with short-range radio sets so that their occupants could talk to each other and to the parent ship. If they were required only for use in space, they might consist simply of cylinders with flexible tubes and gauntlets for the occupants' arms, and large transparent 'bubbles' at the same end for vision: no movable legs would be necessary and this would much simplify design. For work on a planet or satellite with appreciable gravity, legs would obviously be required, and some ingenuity would be needed to prevent the

internal air-pressure from making the suit rigid so that the unfortunate occupant did not become spread-eagled, unable to bend his limbs. When its evolution is complete, the spacesuit may well be almost as complicated as the spaceship, of which indeed it will be a microcosm.

2. Professor Andrade's book *An Approach to Modern Physics* was written for the general reader, and there is an abundant use of example and analogy. The following extract deals with sound as the vibration of material bodies; from it select the main facts and arrange them in a short paragraph as a form of definition of sound.

The study of sound is really the study of vibrations of material bodies. Light vibrations come to us from the sun across space empty of matter, so that, whatever light may be, it is not a swinging of material substance, in the ordinary sense. Sound, however, requires a solid, or a liquid, or a gas if it is to travel; the man in the moon could see any light signal we might make, but could not hear any shout, however loud. Solids convey sound very well, a fact familiar to the heroes of old adventure stories, who put their ears to the ground to hear the hoof-beats of approaching horsemen. . . .

Turning to liquids, bathers can easily verify that water conveys sound very well, for the beat of a ship's screw can be detected when the ship is still far off by putting the head under water. Elaborate methods were worked out during the First World War for detecting the positions of submarines by the sound of the engine conveyed through the water to special microphones. . . .

As for gases, sound ordinarily reaches our ears through the air, and there are many simple experiments showing that it passes equally well through other gases. The old experiment of a bell, driven electrically or by clockwork, and carefully suspended by thin rubber threads in a jar from which air is then pumped out, shows, on the other hand, that sound cannot pass through a vacuum, for as the pumping proceeds the ringing becomes fainter and fainter, until finally nothing can be heard.

3. One of the dramatic moments in Sir Frank Whittle's autobiography, *Jet: The Story of a Pioneer*, is the testing of his experimental engine. This extract deals with a frightening 'runaway' experienced in the early stages. Convert the descrip-

tion into a brief statement of events, concentrating on what happened mechanically during the test.

A safety 'screen' had been erected around the engine consisting of three rectangular sheets of steel plating about one inch thick, two of them standing vertically and the other placed across the top. This was considered to be very necessary because if a turbine overspeeds it is liable to burst, with disastrous results. Pieces may be hurled through the air at speeds of the order of 1,200 ft. per second—that is, almost as fast as a rifle bullet. There had been several occasions in the past when steam turbines had burst with fatal results.

I was at the controls.

When all was ready the fuel pump (driven by a separate electric motor) was switched on and one of the test hands engaged the starter coupling (which was designed to disengage as soon as the main rotor of the engine overran the starting motor). I then gave the signal for starting. When the starter motor had raised the speed about 1,000 r.p.m., I opened the control valve which admitted fuel to a 'pilot' burner in the combustion chamber and rapidly turned the handle of the hand magneto to ignite the finely atomized spray of fuel which this burner emitted. An observer peering through a quartz observation window in the combustion chamber gave me a 'thumbs up' sign to show that the pilot flame was alight. I signalled for an increase of speed of the starter motor, and as the tachometer indicated 2,000 r.p.m. I gradually opened the main control valve. For a second or two the speed of the engine increased slowly and then, with a rising shriek like an air-raid siren, the speed began to rise rapidly, and large patches of red heat became visible on the combustion chamber casing. The engine was obviously out of control. The B.T.-H. personnel, realizing what this meant, took to their heels in varying directions, one or two of them into large steam turbine exhaust casings which were standing near by, and which made useful shelters for such emergencies. I screwed down the control valve at once but this had no immediate effect and the speed continued to rise. Fortunately, the acceleration ceased at about 8,000 r.p.m., and slowly the revs. dropped again.

This incident did not do my nervous system any good at all. I have rarely been so frightened.

The next evening, after an alteration to the fuel system which we supposed would improve matters, we tried again. This time things were even more alarming. The engine accelerated out of control from about 1,500 r.p.m. without the main fuel control being opened at all. Though I switched off the fuel pump thereby cutting off all fuel supplies, the speed rose rapidly. In addition sheets of flame belched from the jet pipe, and clouds of fuel vapour jetting from leaking joints were ignited by patches of red heat on the combustion chamber. Flames were leaping and dancing in mid-air above the engine. The B.T.-H. personnel, alerted by the earlier experience, disappeared even more rapidly. Fortunately, once more the uncontrolled acceleration did not take the engine byond 8,000 r.p.m.

Later that evening Laidlaw, who was staying at the same hotel in Rugby, insisted on my drinking considerable quantities of red wine to calm my nerves after this alarming experience.

There was a simple explanation. Owing to an oversight in the arrangement of the fuel system, it was possible for the main burner to inject fuel into the combustion chamber every time the fuel pump was switched on. Though this injection only happened for a few seconds at a time, fuel pump tests were frequent, and so the amount of fuel unknowingly injected into the combustion chamber was considerable. The shape of the combustion chamber permitted this fuel to accumulate below the fuel burner. It was this 'lake' of fuel which was responsible for the 'runaways.'

4. Read this excerpt from an article by Paul Bennet (*Business*, May, 1956) in which he explains the science of *ergonomics*. Select the important information and arrange it in a compact paragraph suitable as an entry in a general encylopaedia.

Despite its extraordinary name, ergonomics is not just something which has been dreamed up by boffins to confuse busy business men. It is a practical science which can make and is making a valuable contribution to industry. Research being carried out in this field can increase efficiency, improve product design, and make employees more contented.

Broadly, ergonomics is the study of the relationship between man and his working environment. It involves designing or redesigning machines and equipment so that due regard is given to the capabilities and limitations, physical and psychological, of the human beings who have to use or operate them. In contrast to the more common, traditional policy of 'fitting the man to the machine' by selecting and training the best man for each particular job, the ergonomist endeavours to fit the machine to the man—any man.

He accepts people, in fact, as he finds them, with all their quirks and idiosyncracies, and tries to establish their capabilities as machine operators. He then attempts to allocate to the machine and to the human operator the functions each can perform best. His object is to enable a person of ordinary abilities to carry out his tasks safely and efficiently, with a minimum of instruction.

The ergonomist himself needs a knowledge of a number of sciences and techniques in order to carry out his work. These include psychology, the study of human behaviour; physiology, the study of how the human body works; and anatomy, the study of the structure of the human body. In addition, of course, he must know something about engineering and work-study techniques.

Although the word 'ergonomics' was born as recently as 1949, and the concentrated study of this new science dates back scarcely fifteen years, its origin can really be attributed to early nineteenth-century astronomers. They started the ball rolling when they discovered that, contrary to general belief, human reactions were not instantaneous. Observers engaged in measuring the speed of star movements across the lenses of their telescopes did not agree in the times they reported. So a new factor had to be taken into consideration, the time taken by the human mind to react to a message from the eyes.

During the course of the nineteenth-century, physiologists developed various techniques for measuring human reaction time. Research was then taken over by the new science of experimental psychology and knowledge was steadily but slowly accumulated up to the start of World War II.

It was then that ergonomics received its greatest impetus. Scientists were forced to leave the academic seclusion of their laboratories to tackle urgent practical problems. They were called in because designers had created machinery and equipment which surpassed, in terms of accuracy and speed, the capabilities of human operators. Engineering progress had reached the stage where man, and not the machine, was the weaker link in the man-machine system.

So the psychologists, physiologists, and anatomists who had previously worked only in universities and hospitals were called in to advise engineers on the design of weapons and equipment. Among the developments on which they worked was radar, which required the operator to see a small spot (or shape) of light, the Asdic system, which required him to hear a weak noise, and various types of new guns and aircraft which required him to act very quickly.

It was the wartime co-operation between the different kinds of scientists and practical designers which resulted in building up the new specialist knowledge called 'ergonomics'. But just as this new science played an important part in the Allied war effort, so it can make a valuable peace-time contribution to industry in the design of both consumer products and industrial equipment.

5. Leslie Halward, in *Famous British Engineers*, gives an account of the building of Telford's suspension bridge over the Menai Straits. Read the following extract for facts and figures about the building, then express the information in a clearly composed illustration accompanied by a short commentary. Your aim should be to display the information so that it can be read quickly and easily.

The site chosen on the Caernarvonshire side was again Ynys-y-moch, the rocky shore there forming a good natural foundation. The supporting pyramids were to be a hundred and fifty-three feet above the level of the roadway and the distance between the two five hundred and fifty feet. There were to be sixty main chains, with a drop of thirty-seven feet, each chain made up of thirty-six bars of half-inch square iron making it four inches in diameter, the whole welded and braced together with bucklings and iron wire and secured to a mass of stone work, arches, and piers on opposite shores. There was to be a twelve foot carriageway on either side of the bridge and a footpath four feet wide running down the middle. Immediately the Act was passed, Telford went to Bangor to make the necessary preparations for the work.

First it was necessary to blast away some of the rock at Ynys-y-moch to obtain a level foundation for the main pier. A landing quay was erected and connected to the shore by an embankment along which ran railway lines. Stone required for the building was loaded on to sledges and pulled along this embankment by horses, it having been brought by barges from the quarries opened at Penmon Point. The first stone of the pier was laid by the resident engineer, W. H. Provis, without ceremony, on 19th August, 1819.

Each of the main piers, the one in Caernarvonshire and the other on the Isle of Anglesey, rose to a height of a hundred and fifty-three feet, and it was on them that the massive chains, each weighing twenty-three and a half tons, were to be suspended. All was ready for the hoisting of the first of them on 26 April, 1825, almost six years after the building of the piers was begun. A great crowd gathered, and Telford himself was there to superintend the operation. About half-past two in the afternoon, an hour or so before high water, the chain was cast off on a raft from the Caernarvon side and was towed by five boats until it stretched loosely from shore to shore. One end was bolted to another chain hanging ready down the face of the Caernarvon pier, while the other was lashed to ropes fixed to capstans on the Anglesey side, the ropes passing over the top of the pier by means of pulley blocks. Two capstans were ready for actually hauling the chain and each was manned by about a hundred and fifty labourers.

At a given signal a band of pipers struck up a tune and the men, moving at first at a trot, began to turn the capstans. The pace slackened as the weight of the massive chain was taken. Gradually the ends rose; the middle was lifted clear of the raft, which floated away with the tide; and in just over one and a half hours after the start of the operation each end was secured to the satisfaction of Telford, who went to the top of the piers to examine the fastenings. Cheering broke out on both sides, and three excited workmen scrambled along the chain, which covered, because of the curvature, five hundred and ninety feet, presumably to be able to boast that they were the first ever to make the crossing. Telford himself was excited in a different way; it was when friends went to congratulate him on the safe suspending of the first chain that they found him praying.

Telford's method of securing the land ends of the chains was ingenious and proved to be extremely effective. Narrow tunnels were made by blasting the rock, the tunnels gradually sloping down to a final depth of about twenty yards, where, in a sort of compartment properly drained and big enough for workmen to move about in, the chains were held by iron work placed in such a way that they could not break loose, at that point at any rate, without tearing up thousands of tons of solid rock.

6. On a Saturday evening in 1930, the airship R.101 left London on an experimental voyage to Karachi. She was to travel by way of Paris, Toulouse, and Narbonne to the Mediterranean, but as she passed over Beauvais, at about 2 a.m. on 5th October, the huge craft met heavy rain and strong winds. Soon after, she dived steeply into the ground and burst into flames.

Only eight of the fifty-four persons on board survived, and the disaster remains one of the most tragic moments in the history of aviation.

In these days of jet propulsion, that great bulk of gas-filled fabric moving slowly across the sky seems as antediluvian as the pterodactyl, yet only twenty-five years separate R.101 and her sister airships from the 1,000 m.p.h. jet aircraft and the rockets that carry satellites into space.

The following passage is an excerpt from *The Times*, giving a vivid description of the wreckage of the R.101 as seen by one of the reporters. Make a summary of this description, setting it out in enumerated paragraphs or under section headings, restricting yourself to a bare and clinical observation of the wreckage and its position, as though for the benefit of an official examiner or inspector.

> I found the wreck of R.101 near the village of Allonne, a few miles south of Beauvais. As I approached the scene of the accident through the Bois Seck I caught sight of what looked at first like the half-built roof of a railway station standing up on the horizon beyond the trees. This was the framework of the unhappy airship. Getting nearer still, a delicate network of girders could be made out against the sky. It was one of the vertical fins of the ship. All the fabric had disappeared, and as one began to realize the immense scale of the structure, which from a distance of a quarter of a mile already towered above the surrounding trees, it was difficult to believe that these were the remains of an aircraft fallen by accident into this windswept landscape, and not the framework of some gigantic building intended to stand permanently in this place.
>
> With a near view of the wreck the impression of an orderly framework set on the ground gave way to one of tragic chaos. The tail of the airship alone retained recognizable shape, its structure not having been very greatly damaged. It lay at the foot of the slope resting on the lower vertical rudder, which had been torn from its fastenings. The pointed tailpiece, with the framework of an observer's cockpit built into the end of it, towered sixty feet above the ground. From the framework of this part of the ship rags and strips of fabric hung fluttering in the wind. The two elevator planes were still almost covered with fabric.
>
> Moving forward from the tail, the main girders of the airship lay parallel along the ground or hung high in the air from the distorted frames. At first the structure rose into the air from something like two-thirds of its original height, but towards the centre of the ship it sank lower, and dissolved into a shapeless mass. Here in the centre section the fire was fiercest; here, also, the ship must have broken her back when she struck the ground. The stern

engine stood upright on the centre line of the ship. The central gangway and its railings could still be seen. All the gasbags and the outer envelope had vanished, but through and over the distorted framework, wire stays and gasbag nettings made a fine and elaborate tracery as though an army of gigantic spiders had already covered the useless skeleton of R.101 with their webs.

At intervals, curious circular objects, each looking like an aluminium motor tyre with a soup plate attached to it, lay on the ground or hung in mid-air by their wires. These were the great gas-valves by which the buoyancy of the airship was controlled. Near the very centre of the ship a denser mass of wreckage indicated the quarters of the ill-fated passengers. Here domestic fittings, such as staircases and water-tanks, could be distinguished, and on the starboard side a row of gilded pillars—all that remained of the promenade deck—seemed to grin garishly like false teeth in a skull. A few flames still flickered on the ground, twelve hours after the wreck. Into this part of the wreckage men had cut their way with a blow-lamp, chisel and hack-saw to look for possible survivors and to get at the bodies of the dead.

The ridge on which the airship struck is thinly wooded with hazel and oak bushes and among these the forward portion lies. A polished aluminium engine gondola lay among the bushes almost intact, though its base was crushed by the fall. From the stern of the gondola the cylinders of one of the heavy oil engines projected into the air. Behind it the boss of a wooden propeller rested on the ground, the blades, snapped off short, appearing as thick as a young tree. At the forward end the smaller airscrew for driving the electric generator, with the tips of its blades broken, was still in place. On the top of the car lay the cooling radiator; beside it a fuel tank; above it, beyond a few feet of tangled girders, was the open sky.

The extreme nose of the airship lay on the summit of a hill. From this point, looking back down the hill, the whole carcass of the airship lay before me, a mute witness to the tragedy of hopes shattered and lives sacrificed. The long, twisted girders writhed on the ground in parallel lines. On each side the remains of the cross-frames rose like broken ribs, and in the distance over the arched skeleton of the airship's tail, itself as large as the entrance to an ordinary railway station, the frame of her vertical fin rose mournfully against the sky and the balance rudder, still rocking on its hinges, swung aimlessly like an indifferent weathercock in the wind.

7. The following description of the principles used by geologists in searching for oil is taken from a booklet produced by the British Petroleum Company. Read the extract carefully and note that the principles are stated first, followed by an explanation of them in more detail and a brief concluding description of the work of the oil geologist.

You are required to present the information in a shorter, more quickly appreciated form. Use diagrams if necessary, but the minimum of commentary.

> There are four fundamental principles on which the geological basis of oil finding depends. First, the search must be in the area where a considerable thickness of rocks which have been formed in a prehistoric sea or estuary are present; second, these rocks must contain porous or fractured components capable of holding fluids; thirdly, the rocks, originally nearly flat, must have been bent up by earth pressures into a suitable attitude for accumulating any oil which may have been formed; fourthly, the porous or fractured rocks must be overlain by some non-porous rock to serve as a seal.
>
> The earliest clear statement of these principles was made by T. Sterry Hunt of the Geological Survey of Canada in 1865, and it is remarkable that, in spite of all later scientific work, based on enormously improved knowledge of the surface geology of the world and the results of tens of thousands of deep bore holes, Sterry Hunt's early ideas still remain sound.
>
> The rocks of interest to the oil explorer are those called the sedimentary rocks, formed from the mud, sand, or shell banks which for millions of years accumulated on the sea beds of prehistoric seas in areas where the earth's crust was sinking slowly. Most of this material has come from the wearing down of neighbouring lands by the action of frost, heat, wind, and rain, but a proportion of it consists of the hard parts of microscopic organisms which continually rain on the sea bed when the parent organism dies. As this material accumulates, it gradually compacts and hardens, the muds turning into clays and shales, the sand into sandstones, and the shell banks into limestones.
>
> It is now generally accepted that oil is formed from organic matter of animal or vegetable origin, which is deposited with the mud, sand, or shell banks which form the raw material of the rocks. The proportion of original organic matter to rock material probably small, usually less than 2 per cent, but even if only a fraction of this 2 per cent is ultimately converted to the mixture

of hydrocarbons which we call crude oil, the enormous bulk of sedimentary rock available can provide very large quantities of oil. For example, every one-tenth part of 2 per cent of organic matter present in one cubic mile of rock, could produce about 7 million tons of oil, if directly converted to oil, and one cubic mile is a very small volume of rock compared with the amount available in the immediate vicinity of most oil-fields.

The disseminated organic matter, once safely entombed beyond the reach of atmospheric oxygen, is gradually changed by chemical, physical and bacteriological action into minute drops of crude oil which float in the salt water filling the remaining space between the grains of rock matter. As the rock material compacts, the salt water and the oil droplets will gradually be squeezed out of it via the easiest paths of escape, which will be through the porous sandstones or limestones. It is during the movement of the fluids through the rocks that the oil droplets begin accumulating in any porous or fissured sand or limestone layer that may be available. This accumulation is partly due to the tendency of the oil to float upwards through the water and partly to other physical processes. It is the floating upwards which causes the oil to be finally trapped in suitable rock structures. All these processes, of course, take place very slowly, the rate of oil and water movement being probably only a foot or two a year, the period of accumulation of an oilfield being measured in millions of years.

An important part of the process of oil accumulation is played by the mountain-building movements which periodically affect the earth's crust. It is now well known that these mountain-building movements are particularly prone to occur where the earth's crust has been sinking gradually for millions of years, and where the hollows so formed have been gradually filling up with sedimentary rock. It is in these very areas that our oil droplets have been forming and it is the mountain-building movements which form the traps which will subsequently accumulate the oil. The oil geologist, therefore, having satisfied himself that an area fulfils the primary conditions of an oil prospect, which he will do by an examination of all the rocks visible at the surface, or penetrated by previous bore holes, must map the attitude of the rocks to see whether suitable traps are present. The anticline, or dome of rock, is the most simple form of structural trap. One of the main tasks in oil exploration, therefore, is to make a three dimensions map of the rocks, and for this work the geologist and his colleague the geophysicist use whatever tools are most appropriate for the particular area under examination. The complexity of the tools has increased enormously since the oil industry began.

8. The following extract is from an article by Sir Graham Sutton, F.R.S., published in the *New Scientist*, on the subject of tornadoes. After reading the passage carefully, select the principal facts about the tornado and express them concisely in your own words using the following framework:

Formation of tornado.
Characteristics of tornado.
Effects of tornado.

A tornado should not be confused with the tropical hurricane, or typhoon, which consists of a mass of air a few hundred miles in diameter circulating at high speed, often more than 100 miles an hour, around a central calm region known as the 'eye' of the storm. The tornado is rarely more than 300 yards in diameter and often much less, and its life is usually reckoned in minutes, whereas the hurricane lasts for days. The appearance of the tornado is familiar from photographs, which shows a dark writhing funnel extending from a low black cloud to the ground. The entire force of the wind is confined to this funnel, and when a tornado strikes a built-up area it cuts a narrow swath of destruction, leaving areas on both sides of the path completely untouched.

The outstanding feature of the tornado is the tremendous speed of the swirling air in the vortex. There are virtually no accurate records of the velocity of the wind because any anemometer which by chance was in the path of the storm is certain to be destroyed, but estimates have been made from the damage caused, and from characteristic effects, such as pieces of straw driven into trees. These estimates suggest speeds in the neighbourhood of 300 miles an hour, and it is possible that 500 miles an hour has been reached in some very violent storms. Such speeds are, of course, far in excess of those found in other atmospheric disturbances, and in this respect the tornado is unique.

The rapidly rotating air is accompanied by a very sudden large drop in atmospheric pressure. Here again there are few reliable records, but a fall of 200 millibars (nearly 6 inches of mercury) was reported from Minnesota in 1904. Such tremendous evacuation of air causes buildings in the path of the storm to explode outwards, and this type of damage is typical of the tornado's trail of destruction. The track of the tornado is usually short—often not more than about ten or twenty miles—and its speed of movement varies between twenty and fifty miles an hour. Sòme extraordinary effects are reported from time to time, especially those caused by the immense updraught at the centre of the funnel. There are well-authenticated accounts of wooden buildings, cars and animals being picked up bodily and deposited many miles away, and there is one story, from Kansas, of a herd of cows emulating the nursery rhyme and taking off for the Moon, 'looking like gigantic birds in the sky.' Showers of frogs, whipped up from their native ponds and carried aloft for miles, are not uncommon.

Tornadoes have been reported from all inhabited parts of the world, but they occur most frequently in the United States and in Australia. Despite popular opinion, they are not confined to the warm season. An examination of U.S. Weather Bureau records made by Fawbush, Miller and Starrett shows that for the United States as a whole, tornadoes were reported during every month of the year in the period 1920–1949. The incidence of tornadoes was greatest between March and June, with May as the peak month. . . . In the Middle West the peak period is in early summer. The most unlucky state appears to be Kansas, with a total of 587 tornadoes between 1915 and 1949, but Iowa runs a close second with 512 in the same period. . . .

From the aspect of meteorology, the tornado approaches more closely to the 'vortex' of theoretical hydrodynamics than any other atmospheric disturbance. The system resembles in many ways the little whirlpool seen at the drainage hole of a bath. The mechanism of the tornado is still obscure, but its formation is clearly associated with instability in the lower part of the atmosphere. The storms form in the vicinity of strong cold fronts and squall lines, and often accompany thunderstorms. The essential initial condition appears to be the presence of a deep layer of relatively warm moist air near the ground, with a layer of cold dry air above. In such conditions convection currents form easily, but it is not known what precise combination of circumstances causes the typical narrow pendant vortex to form. There is some evidence that the funnel originates in a horizontal vortex in the parent cloud, whose intensely black colour is evidence of tremendous condensation of water-vapour and therefore of the release of large amounts of energy in the form of latent heat. There are, however, many difficult problems, such as the maintenance of the very low pressure at the centre of the vortex despite the strong inflow at the surface, which still awaits a satisfactory quantitative solution.

9. One of the great feats of civil engineering was the construction of the railway tunnel at Box, on the London-Bristol line. This excerpt from L. T. C.Rolt's biography of Isambard Brunel, gives a description of the project. Reduce the original to the form and size of a short newspaper article, making sure that you retain and emphasize the facts, particularly the statistics.

By the end of May 1841 there remained only one missing link in Brunel's iron chain from London to Bristol and it was the hardest of all to forge. This was the section between Chippenham and Bath. It included much deep cutting and embankment, yet another crossing of the Avon at Bath, viaducts at Chippenham and Bath, and the diversion of the Kennet and Avon canal, but, over and above all, there was that "monstrous and extraordinary, most dangerous and impracticable tunnel at Box." Nearly two miles long, it was by far the greatest railway tunnel so far attempted, exceeding the length of Kilsby on the London and Birmingham Railway by 768 yards. Apart from its length the fact that Brunel had planned it upon an incline of 1 in 100 provoked . . . much headshaking amongst the pundits.

Work on the tunnel had begun as early as September 1836 with the sinking of six permanent and two temporary shafts 28 feet in diameter from the hill top down to rail level. One of these, sunk near the road from Bradford to Colerne, was 300 feet deep. All had been completed by the autumn of 1837 and contracts for the tunnel were then advertised. George Burge of Herne Bay, one of the first of the great railway contractors, undertook to build over three-quarters of the length of the tunnel. His section ran through clay, blue marl, and inferior oolite, and was to be lined with brick throughout. The remaining half-mile at the eastern end was undertaken by two local men, Brewer of Box and Lewis of Bath. It was to be driven through the great oolite and Brunel planned to cut it in the form of a Gothic arch and to leave it unlined. One of his personal assistants, William Glennie, remained in charge of the whole work until its completion.

It was an immense undertaking and the more staggering when we remembered that, apart from the steam pumps which kept the workings clear of water and the power of gunpowder which was used to blast away the rock, it was accomplished entirely by the strength of men and horses working by candlelight. For two and a half years the work consumed a ton of gunpowder and a ton of candles every week. The lining bricks were burnt by another local man, Hunt, at a yard in the meadows to the west of Chippenham and for three years he employed one hundred horses and carts to convey a total of 30,000,000 bricks to the site.

The task undertaken by Lewis and Brewer in driving the great gothic arch through half a mile of solid Bath stone was almost superhuman and was made the more difficult by the inrush of water through fissures, particularly during the wet winters. In November 1837 there was such an overwhelming influx that the water overpowered the pumps, filled the tunnel and rose 56 feet up the shafts. The situation was mastered by the installation of a second pumping engine of 50 h.p., but another inundation in the following November held up work for ten days.

The soil excavated, 247,000 cubic yards in all, was hauled up the shafts in buckets by means of horse gins at the surface. It is entirely typical of Brunel that when he was accompanied by some of his directors on one of his many visitors to Box he should have insisted they join him on his descent of one of the shafts. . . . Lewis and Brewer had commenced cutting their section from opposite ends and Brunel was present in the workings on the historic occasion when the two gangs at last met. Whether or not a true line had been kept was a question which had been the subject of much anxiety, and when it was found that the junction was perfectly true Brunel was so delighted that he took a ring from his finger and gave it to the foreman in charge of the gang. This ring is treasured by his descendants in Bristol to this day.

August 1840 was the date which had originally been stipulated for the completion of the tunnel contract, but despite every effort the work fell far behind schedule until, in December 1840, Brunel decided to throw all his reserves into this last battle. Henceforward a force of 4,000 and 300 horses laboured night and day to complete the tunnel. . . .

In June 1841, the great tunnel was at last completed and the way opened throughout from London to Bristol. The whole work had cost £6,500,000, considerably more than double Brunel's original estimate, but the dream of the *Bristol Railway* which had seemed to many so impossible of realization only eight years before was now a magnificent reality. On the last day of the month a decorated train pulled out of Paddington and arrived in Bristol four hours later.

10. The following excerpts from A. E. McKenzie's *The Major Achievements of Science* deal with the life and career of Michael Faraday. Use the information to compose a short

biographical note, similar to the type given in a reference work, or in *Who's Who*.

Michael Faraday (1791–1867), the third child of a London blacksmith, was brought up in poverty, with no formal education beyond the three R's, and became one of the greatest of experimental investigators. He was, however, extremely fortunate in his opportunities for further education. He began his working life as errand boy to a bookseller, and then was taken as an apprentice without premium. Inside the bookshop he came upon Mrs Marcet's *Conversation on Chemistry* and an article on electricity in the *Encyclopaedia Britannica*; these gave him his first interest in science. He was impressed also with Watt's *Improvement of the Mind*, and he made serious and sustained efforts to teach himself to speak and to write correctly. He took lessons in drawing from a French *émigré*.

One of the customers of the bookshop took Faraday to hear Davy lecture at the Royal Institution. Faraday sat in the gallery and was entranced. He made copious notes, and wrote up a full account of the lectures, with diagrams, which he sent to Davy, begging for an opportunity of working at the Royal Institution. He said that trade was 'vicious and selfish' and later confided to Davy, who must have listened with an indulgent smile, that he thought that men of science were animated by higher moral feelings than other men.

Davy was kind and sympathetic, but told him that science was a hard mistress and that he would do far better in bookselling. It so happened, however, that shortly afterwards the laboratory assistant at the Royal Institution was dismissed for misconduct, and Davy offered Faraday the post, which was accepted with alacrity.

Seven months after his appointment at the Royal Institution, in October 1813, Faraday set out with Davy, as secretary and valet, on a tour of France, Italy, Switzerland, Germany, Holland and Belgium, lasting one and a half years, during which he met most of the leading men of science in Europe. He was at that impressionable and receptive age at which some young men are university students. He had never before travelled more than twelve miles from London. He was excited by the journey to Plymouth; in Paris he had a glimpse of Napoleon in his carriage; he took part in experiments on electric fish at Genoa he visited the laboratory of the *Accademia del Cimento* at Florence, and saw Galileo's first telescope; in Milan he met Volta, then an old man; he climbed Vesuvius; he accompanied Davy on fishing expeditions in the Alpine streams and acted as loader when Davy went shooting with de la Rive near Geneva; on his way home he slept in Brussels, two months before Waterloo. The travellers wandered through France and Italy without molestation during the Napoleonic War. . . .

Back in London Faraday spent seven years as Davy's assistant. In Copenhagen in 1820 Oersted discovered that an electric current flowing in a wire caused a nearby magnetic needle to be deflected. This stimulated intense interest in electricity and magnetism, and in 1821, Faraday wrote *Historical Sketch of Magnetism* which gave him a mastery of the work of other men in this field. In the same year he made his own first important contribution. He caused a magnet to rotate round a wire carrying a current, and also a current-carrying wire to rotate round a magnet—the first primitive electric motors.

In 1823 Faraday was proposed for election as a Fellow of the Royal Society; Davy, who was President of the Society, tried to have his candidature withdrawn. Davy was a sick man at the time, but he was also jealous. Never again were the relations between the two men the same. Faraday was elected a Fellow and two years later became Director of the Royal Institution. . . .

Faraday married in 1819 and took his wife to live in a flat over the Royal Institution, where they remained for over forty years. During this time he performed thousands of experiments which are recorded in his diary, now published in seven large volumes. He was always full of ideas and used to say that he was content if one in a thousand of his experiments led to a really important discovery. . . .

Faraday's pleasures were simple. Occasionally in the evening he would take his wife to see Kean act or to hear Jenny Lind sing. He loved to watch the antics of the monkeys at the Zoo. . . . He sometimes rode his velocipede to Hampstead in the morning, or round the lecture theatre of the Royal Institution after the day's work was done. . . .

The theatre held 700 persons and among its audience were sometimes to be found the Prince Consort and the Prince of Wales. Charles Dickens, then editor of *Household Words*, asked Faraday if he could report the lectures in his magazine. Farady initiated the Christmas lectures for children, and the Friday evening discourses.

Faraday's salary, when he was married, was £100 per year with house, coals, and candles. By 1831 he was making £1,000 a year as a consultant scientist, and Tyndall reckons that he could have raised this to £5,000 a year. He found, however, that too much of his time was taken by these activities, and he deliberately sacrificed wealth to scientific research. In 1832 his income fell to £150 a year.

Faraday found it necessary deliberately to restrict his interests as well as his finances, in order to pursue his all-absorbing investigations. At the age of 49 he had a mental breakdown and suffered from giddiness and loss of memory, which necessitated a four years' rest.

One outside activity Faraday permitted himself. He became scientific adviser to Trinity House. . . . His visits of inspection to the lighthouses all round the coast provided him with a much needed change and relaxation.

In 1858 Queen Victoria gave him a grace and favour house at Hampton Court. Here he spent his last few years sadly affected by loss of memory.

11. The following extract, from Charles-Noël Martin's *The Thirteen Steps to the Atom*, contains a description in some detail of the operation and advantages of the electron microscope. Prepare an outline of the main information, suitably and interestingly expressed, so that it can be used on a display card when exhibiting the microscope. Use sub-headings if you wish.

In 1932 the German Ruska constructed the first *electron microscope*. This was a practical application of the theoretical discovery (eight years earlier by the Frenchman Louis de Broglie) that electrons, which until then had been thought of as electrical corpuscles, behaved in some circumstances as if they had a waved character. The experimental confirmation of this came in 1927, when the Americans Davisson and Germer diffracted electrons from a crystal of nickel.

The superiority of the electron miscroscope over the optical instrument is essentially a consequence of the fact that the wave-length associated with a beam of fast electrons is less than that of light. For this reason the detail which can be distinguished by an electron beam is much smaller than that accessible to an optical microscope. Though, in principle, use might be made of beams of other kinds of atomic particles (which also have a wave-like character) for microscopy, it turns out that electrons can most conveniently be used to fill the gap between atomic dimensions and the lower limit of resolution of the optical microscope. The wave-length associated with a beam of protons, on the other hand, would be so small that the pictures taken in this way would be confused by the presence of effects due to the structure of atoms. It is interesting, however, to note that beams of neutrons have been used to supplement the use of X-rays in the exploration of crystalline structure.

Though these are the considerations which determine the ultimate power of the electron microscope, in practice its method of operation may be pictured more simply. In all but special circumstances an electron microscope must be used with specimens which are sufficiently thin to allow electrons to pass through them with some ease. If some parts of a thin specimen are thicker than others, or if they are more dense, and thus more able to stop electrons, there will be variations in the proportion of an electron beam transmitted through different parts of the specimen. In an actual instrument arrangements are made to fire a beam of electrons at such a thin specimen. A photographic plate or a fluorescent screen placed behind the specimen can then record these variations. The magnification of the specimen in the record depends on the geometry of the beam of electrons.

The wave-like character of electrons is a limit of the fineness of detail which can be distinguished in this way, because in its passage through the specimen a particular electron is affected by the nature and thickness of the material within a wave-length or so of its path.

At present there are about 2,500 electron microscopes at work in the world. The most advanced is the Elmiscop, developed by Ruska in Berlin. Japan has the highest output of these instruments, with four firms building and exporting them. Russian production is estimated to have been between five hundred and a thousand.

Though some research workers have succeed in obtaining photographs of specimens by means of the reflection of beams of electrons from their surface, for all practical purposes extremely thin specimens must be examined by transmission. The preparation of these specimens is an art in itself.

The most direct method, but the most recently developed and perhaps the most difficult, is to cut thin sections of, say, a large biological object. Methods of doing this which yield specimens about 100 ångströms thick are now available, and can conveniently be used for studying the structure of wood and so on. The interpretation of the photographs obtained in this way is not, however, as simple as it might seem to be: for one thing, a very thin slice of biological material will rapidly lose the water it may contain when it is mounted in the vacuum inside an electron microscope.

Objects which are themselves extremely small may be examined by mounting them on a thin film of material which is itself transparent to electrons: collodion is often used for this purpose. Small crystals, spherules of rubber latex, and so on may be examined in this way. Though this technique is mainly suited to the study of the shapes of very small objects, further details of surface relief may be thrown up by coating the specimen with a material which is opaque to electrons. If the coating is done by projecting a beam of the opaque molecules at an oblique direction on to the specimen the result will be that in the electron microscope some of its structure will be revealed by the 'shadows' cast on the uniform background.

To examine the structure of large specimens, however, other techniques are needed. Usually an attempt is made to construct a replica of the surface of a specimen which will itself be thin enough to transmit the electrons that would be stopped by the original. Thus the surface of a crystal may be examind by coating it with a thin layer of plastic material which, when stripped off, will carry with it a representation of the crystal surface. For technical reasons it is often necessary to make a replica of the structure of this first replica, and in the process it is, of course, essential that distortions of the structural detail be made as small as possible.

VIII

TECHNICAL VOCABULARY

WORDS

Words are basic units of expression. They are the labels of your experience and it is important that you use the correct labels if you wish to explain your ideas, opinions, and observations successfully to other people.

A sound knowledge of words is essential if you intend to read and write intelligently; but this does not mean merely knowing how to spell and having a vague idea of word meaning. To use a word properly you should know how it has been formed and the meaning it has developed in certain situations and in company with other words. Your ideas will be put across more accurately if you can choose the right word to fit your expression.

Word accuracy is particularly important in exposition, where the vocabulary has developed from the need to express ideas and information objectively. You must know how to differentiate between words with similar meanings and you must never be content to say, "This word will do; it says roughly what I mean." The word should say *precisely* what is intended.

Skill in the recognition of words comes with wide reading and with much written practice; it also develops from the intelligent use of a good dictionary. Do not be too lazy or too self-conscious to use a dictionary.

Now that different specializations tend to overlap, there is an accompanying overlap of vocabulary. Physics and electronics, for example, have a number of words in common; so have civil engineering and geology. Yet there are often different ways in which simple words are used and you must be aware of those differences.

You should keep up to date with new words. The language of science and technology grows daily, and, although it brings a number of 'gimmick' words and certain slang terms and phrases, it needs newly coined words or old words adapted to fit the new techniques. Do not underestimate the difficulty and complexity of the vocabulary of our age. Remember that, as technicians or technologists, you have two basic languages—Mathematics and English—in which to express your ideas, experiments, results, reports, or opinions. Each language serves a specific purpose and neither must be neglected. Use words with the same care that you would use mathematical symbols and formulae, because sloppy English is as dangerous as careless mathematics.

WORD FORMATION

Many of the words that we use in modern science and technology are of recent date; some go back to the beginnings of science in Ancient Greece or were brought with the caravans from the East. A number of our workshop terms developed in the dialects of this country, invented and used by craftsmen. Whatever the origin, the history of a word widens our appreciation of its meaning, and a knowledge of the way in which scientific words are formed will also help us to fathom the meaning of new words and later to recall and use them in our own work.

Take an interest in the background of the vocabulary that you use. Refer to one of the increasing number of specialist dictionaries that are now available; but do not be content with a narrow vocabulary. Many of the exercises in this book deal with terms from other specializations. Study them carefully, for your own branch of learning is expanding to use not only the equipment but the vocabulary of other technical subjects.

Use of classical terms and affixes

The commonest method of forming scientific and technical terms is to use Latin or Greek roots together with specific prefixes and suffixes. At one time there was a tendency to observe the rule of keeping Latin and Greek formations apart; but the desperate need for new words has meant that the various terms and affixes have intermingled.

Certain affixes have contributed to a large number of different terms and have established themselves as basic units in word formation: auto- -(o)logy -meter sub- bi-. There is a large number of such affixes and in the following series of exercises only a selection of them can be dealt with.

EXERCISES

1. Show how numeral prefixes are used in the following words. Give the meaning of each word.

heptagon	polyhedron	millimetre
centigrade	quadruped	tripod
diameter	monotonous	bicentenary
decennial	kilocycle	tripartite
septet	sextuple	octant
quinquennial	centipede	hectogram
decalitre	octonal	

2. Write down the term given to each of the plane figures from four to ten sides, inclusive.

3. Add root words to each of the following numeral prefixes to form words other than those given or used in the previous exercises. Explain the meaning of each:

bi-	mono-	sept-	deca-	hexa-
penta-	tri-	kilo-	octo-	quadr-

4. Give single words for each of the following short definitions:

(*a*) recurring every eight years
(*b*) consisting of four parts
(*c*) appearing twice in a year
(*d*) solid figure with five faces
(*e*) counting system with ten for a basis
(*f*) a group of three compositions
(*g*) one-sided
(*h*) a distance of ten metres
(*i*) consisting of two algebraical terms
(*j*) destroy a tenth of

5. Add a prefix to each of the following words to convert it to a negative version:

moral	literate	practicable
responsible	possible	theist
credible	legal	ability
appearance	accuracy	metallic
movable	logical	reparable
audible	correct	ambitious
practical	responsible	

6. The following prefixes are used to extend the meaning of root words or stems by such means as reversal of direction, alteration of scope, and position. Show how each prefix can operate, giving examples where possible from your technical vocabulary.

re-	circum-	ultra-	peri-
sub-	dia-	para-	trans-
hyper-	inter-	as-	ambi-
contra-	extra-	de-	ex-
in-	per-	ob-	epi-
hypo-	off-	ante-	intro-
pre-	dis-	se-	syn-

7. Each of the following acts as a prefix or as a component of a compound word. Show how each operates and how it can help to present the meaning of the examples given. Then give further examples of your own.

poly-	polygon	polyhedron
hydro-	hydrogen	hydrometer
micro-	microfilm	microphone
tele-	telephone	telegram
mega-	megaphone	megaton
thermo-	thermometer	thermonuclear
astro-	astronomy	astrology
iso-	isobar	isotope
psych-	psychiatry	psychology
aero-	aeroplane	aeronautical

8. The suffix *-ics* has become the customary form of indicating names of sciences. Show the significance of this in the following words. Explain the science indicated.

thermodynamics	acoustics	physics	economics
mathematics	ergonomics	ballistics	electronics
mechanics	geophysics	optics	statistics

9. This is a more difficult collection of *-ics*. You will need an up-to-date scientific dictionary for these words.

cybernetics	psychometrics	cryogenics
perastadics	semantics	genetics
thermionics	aerostatics	

10. -*(o)logy* is a suffix which indicates a 'branch of knowledge.' Indicate how this suffix helps in the following compound words. Identify each branch of knowledge; then give further examples of your own.

geology	geomorphology	biology	psychology
meteorology	mineralogy	sociology	technology
anthropology	archaeology	pathology	neurology
bacteriology	philology	physiology	zoology

11. This is a list of more difficult -*(o)logy* compounds:

ethnology	agrostology	radiology	virology
cytology	ecology	ontology	embryology
musicology	genecology		

12. The suffix *-scope* represents the Greek *skopos* meaning 'a watcher.' It is used to indicate an instrument for observing. Comment upon the construction and meaning of the following *-scope* compounds:

periscope	telescope	microscope	kaleidoscope
stethoscope	gyroscope	bioscope	scleroscope
tachistoscope	episcope	epidiascope	oscilloscope

13. What is the significance of each of the following prefixes and suffixes when used to form compound words? Give at least one example of each:

-meter	ge-	hygro-	pyro-
-graphy	-ism	stereo-	micro-

14. Explain the way in which the following compound words have been constructed:

pantograph	counterboring	influx
hydrostatics	geography	micrometer
photogram	parachute	monoscope
superheterodyne		

SPECIAL TERMS

A particular science or 'trade' tends to have its special terminology consisting of slang and colloquial terms mixed with a semi-technical vocabulary which probably had its origin in earlier slang and workmen's terms. There are not many 'classical' terms in the engineering workshop. Many of the words have a strong Saxon flavour.

Study the words in the following lists. Make sure you know both the meaning and usage. Try to find time to study their origin as craft words:

1. *Colloquial shop terms*

blow	teemed	wiped	tagged	cope
drag	bosh	swarf	dog	suds (slurry)
snap	clam	shims	crowning	oddside
dwell	sweating	cheeks	poppets	crazing
ribbon	chips	offer		

2. *Radio and Acoustics*

flutter	wow	feedback	tweeter	squawker
scatter	growler	woofer	warble	wash-out

3. *Automobile Engineering*

soup	drag	booster	con-rod	drift
rimbellisher	slap	scuffing	anchors	propshaft

4. *Smithying*

pein	fuller	hardie	swage	necking
flatter	set-hammer	drifts	upsetting	tup
palletts	flash	spelter		

5. *Machine tools*

rake	helix	crowning	shank	arbor
leadscrew	jig	honing	pinion	chatter
cotter	web	trammels	warding	witness
spigot	land	lapping	skew	rack
tolerance	plummer	jennies	apron	reaming
parting	coolant	hob	broach	profile

This is but a small part of the vocabulary of special terms. The fields of civil engineering, aeronautics, electronics, etc., have each developed specialist terminologies. You should be able to move among the words of *your* specialist subject with ease and authority, but you should also be conversant with the terms used by the sciences and technologies that border upon your own.

Supplementary vocabulary exercises are provided at the end of this section. They are divided into specialist groups to provide you with specific practice material.

ACRONYMS

In an age when millions of words are churned out in print every day, and speed is often the governing factor in science and business, there is a natural tendency to look for 'short cuts' in words. This might be termed 'verbal streamlining,' the best example of which is the smoothing of abbreviated forms into a single word.

E.g. A particular piece of equipment is known originally as a *mathematical analyser numerical integrator and computer*

This mouthful can be abbreviated by the simple process of selecting the initial letter of each word:

M A N I A C

This abbreviation can be smoothed, obviously, into the term MANIAC

Similarly:

United Nations Organization
becomes U.N.O. and is smoothed to UNO

This method of formation is known as an *acronym* and can be of two types:

(*a*) Formed from the first letters in each word—as in MANIAC and UNO
(*b*) Formed from parts selected from the main words in a group *e.g.* *Mot*or hot*el* can be reduced to *motel*

Study the following list of acronyms and note carefully how they have been developed:

goon	*g*orilla bab*oon*
BENELUX	*Be*lgium, *Ne*therlands and *Lux*embourg
smog	*sm*oke f*og*
radar	*ra*dio *d*etection *a*nd *r*anging system
GATT	*G*eneral *A*greement on *T*rade and *T*ariffs

EXERCISES

1. In each of the following questions you are given the original term followed by the acronym. Indicate how the acronym was developed.

(*a*) rocket assisted take-off	rato
(*b*) will carry out orders	wilco
(*c*) thermal resistor	thermistor
(*d*) binary automatic computer	binac
(*e*) breakfast and lunch	brunch
(*f*) North Atlantic Treaty Organization	NATO
(*g*) Electronic Random Number Indicator Equipment	ERNIE
(*h*) Analytic computer	anacom
(*i*) Graphite low energy experimental pile	gleep
(*j*) Fog investigation and dispersion operation	FIDO

2. Construct what you consider to be a suitable acronym from each of the following terms. Compare your formation with the acronyms which have already been developed and which are given in *The Dictionary of New Words* by Mary Reifer.

(*a*) United Nations Educational Scientific and Cultural Organization
(*b*) Sound navigation ranging
(*c*) Million electric volts
(*d*) Condensation trail
(*e*) Sound detection and ranging
(*f*) jet-assisted take-off
(*g*) addition product
(*h*) Minimum orbital unmanned satellite of the earth
(*i*) Blind Approach Beacon System

3. Can you trace any acronyms in your own special subject? Are there any familiar ones in your daily newspaper reading? *E.g.*, *Euratom*. Make a note of any such acronyms and try to explain their construction.

NEW WORDS

Acronyms are only one part of the flood of new words since the end of World War II, many of the words horrible in sound and construction. Increased specialization and new scientific and technical operations have helped to spawn such words as:

fluonomist	(U.S. term for a chimney-sweep)
miniaturize	to make smaller, or in miniature,
hospitalize	to put into hospital (noun: hospitalization)
heparin	name for an anti-coagulant discovered in 1946
count-down	Counting the remaining seconds before firing a rocket

Are all these terms really necessary? 'heparin' serves a useful purpose; 'count-down' is a straightforward product of the rocket age; but do we really benefit from a word like 'hospitalize' or an ugly euphemism such as 'fluonomist!'

The '-ize' craze in word construction has produced a host of world such as: 'miniaturize,' 'depersonalize,' 'finalize,' with their '-ization' nouns.

Accompanying this bloating of vocabulary is the jargon of politics and economics: depurge, disinflation, deviationism. Beware of such words. Avoid them if they add nothing significant to your statements; look always for the simple, direct term and use it in preference to the long, pompous word.

Exercise

The following list of words includes some that are *suspect*. Discover the meaning of each word and how it is normally used. Select those words which you consider to be *unnecessary* additions to the language, and give your reasons.

dehumidification	sputnik	fluoridation
desegregate	liquidate	shortfall
bebop	deviationist	phon
cheesecake	hormonate	disincentive
recategorize	streptomycin	coventrate
additive	afterburner	barodynamics
cannibalize	megaton	micronize
admass	disinfest	comparator
rodenticide	automate	subtopia
conurbation	dracone	beatnik
cosmonaut	breathalyzer	bulldozer
coleopter	ergonomics	blip
jeep	aqualung	brainwash

COMPOUND WORDS

Technical terms are sometimes constructed by combining words to act as nouns or adjectives. The great danger with such compounds is that they tend to become over-long and unwieldly and lead to a use of acronyms and abbreviations to overcome this.

Compound words may be:

(*a*) completely united	airframe
(*b*) hyphenated	wind-tunnel
(*c*) accepted as standard phrases	saw bench; air brake

There is no simple rule for hyphenation of compounds, but you should note the distinction between noun and adjective as expressed in the following use and omission of the hyphen:

high frequency (noun) high-frequency (adjective)

If in doubt, consult a good technical dictionary.

Certain phrases may be formed by the use of operative adjectives, and the results may almost be regarded as single terms:

e.g. specific gravity
specific heat
specific volume

CHOOSING THE CORRECT WORD

Word accuracy is important, especially when there are several words with *similar* meanings. You must learn to differentiate between words and to choose the precise word that fits your scheme of ideas. 'Fault' is not the same word as 'error,' but many students will persist in muddling the two words. 'Inclined' is similar to 'slanting,' but the words are not interchangeable and each must be used in its correct context.

There is no simple rule for choosing the correct word; the skill comes from careful reading and familiarity with words in their contexts. Get to know the words really well, then you will know how to use them.

Exercises

1. Explain the meaning and use of each word in the following sets and show how it differs from the others in its set.

edge rim verge lip hem
workman employee operative labourer
cut slash gash slice slit incise
alter change modify reform remodel transform
measure assess rate gauge value estimate
rotate revolve turn spin gyrate
fold crease corrugate crumple
member part section component unit
scrape file abrade grind rasp grate scour
imperfect deficient defective faulty inadequate

2. Use each of the words in the following sets in a sentence to show how it differs from other words in the set.

permeate percolate penetrate
elastic tensile resilient
absorb adsorb imbibe
humid moist wet damp
multifarious multiform multiple manifold
oblique inclined slanting diagonal transverse
efficient effective effectual efficacious
interpret explain translate decode exemplify
strident hoarse discordant gruff grating
pliable flexible plastic ductile malleable

3. A key word has been omitted from each of these sentences. Choose the correct word from the list given below the exercise.

(*a*) In the girder construction, particular difficulty was experienced with the vertical . . .

The . . . of a force F, along a line making an angle θ with the line of action of F, is $F \cos \theta$.

The function of this switch is to connect or disconnect two . . . of the circuit.

A lux (sometimes called a metre candle) is a . . . of illumination.

Although he had not finished his report he submitted the first . . . for consideration.

(part, section, component, member, unit)

(*b*) The workmen in the shop were . . . and found no common expression of their discontent.

The two engines were . . . and only the efficient one remained in gear to drive the propeller.

The wires were first . . . from the terminals.

He . . . third gear and coasted for a while in neutral.

(disconnected, detached, disengaged, disunited, uncoupled, unlinked, unshackled)

(*c*) The inspection department will . . . any work that is skimped or inaccurately finished.

The pump will . . . the waste from the plant into the river at a high rate.

The ram operation is designed to . . . the hot metal through the die in hollow section.

The specially designed equipment enables a pilot to . . . himself from his aircraft in an emergency.

(eject, reject, discharge, expel, extrude)

(*d*) The improvement of the hydraulic system will . . . failures in operation.

We can use a rotary pump to . . . the gas in this cylinder.

The operator is preparing to . . . the load on the machine.

The circulation of cold water will . . . the exhaust steam from the turbine.

(contract, condense, compress, attenuate, reduce, lessen)

(*e*) The vein was . . . with a trace element in an attempt to investigate the source of the trouble.

The fabric was . . . with a fluid designed to repel moths.

A new bonus scheme was . . . into the works in an attempt to step up production.

The machine was completely automatic and no one . . . or removed the material that was being worked.

(introduced, inserted, instilled, injected, impregnated, infused)

ONE WORD, SEVERAL MEANINGS

The different specializations have given to some ordinary words a special significance. A simple word such as 'step' is no longer restricted to the concrete structure outside your front door, or leading down to your lawn. For the aeronautical engineer, a step is part of the construction of the hull of a flying boat, designed to help in take-off. For the sailor, or the civil engineer, the word has a verbal use—to place a mast or a lock gate in a vertical position. For the electronics engineer it can mean a radar pulse superimposed on a time-base trace.

It is important, when you read technical material, to be prepared for such differences. Do not take words always at their face value, nor at the value *you* place on them. Consider the author and his background. What does he intend the word to convey?

If you use any such word in your writing, consider your reader and his chances of misinterpreting the term. If you think he will go astray, define the term for him.

EXERCISES

1. The following words have a variety of uses according to the different specializations shown. Use each word in different sentences to show these uses.

trap	plumbing; geology; aeronautics
screen	photography; radio; textiles; psychology
pulse	medicine; electronics; aeronautics
gate	metallurgy; electronics; mechanical engineering; photography
camber	civil engineering; aeronautics; mechanical engineering
flux	metallurgy; physics; mathematics
grid	civil engineering; electrical engineering; mechanical engineering
battery	electrical engineering; military science; psychology; agriculture
drift	aeronautics; civil engineering; meteorology; geology; mechanical engineering
frame	aeronautics; photography; civil engineering; electronics

2. Identify the different specializations which have given a distinctive meaning or usage to each of the following words:

marrying	crab	truck	vector
turret	pack	load	ceiling
element	modulate	activate	boom
accelerator	absorption	core	threshold
damper	apron	bleed	condenser
pitch	bank	feed	key
rake	pad		

DEFINITION

We define *words* or *terms* not things. Defining a word involves giving its limits as a sign or symbol, usually by explaining to a listener or reader how we are using the word and what we want them to make of it. How we define a word depends largely upon our immediate purpose.

Definitions are not fixed indications of meaning, but brief explanations of the way a word is to be used, so that there will be no misunderstanding or misinterpretation. We can say:

"When I use the word energy I mean. . . ." (or "I want you to take it to mean. . . .")

"When Dr Smith uses the word energy he means. . . ."

"Energy is the. . . ."

We may define a word or term:

(*a*) *By example*—Pointing to, or otherwise indicating the object we mean by the word.

(*b*) *By synonym or antonym*—Defining a word by comparing it with a similar word or by saying that it is the opposite of another.

e.g. "By energy I mean the physical attribute that is the reverse of lethargy."

(*c*) *By analytic definition*—Using a scientific classification to produce a definition formula:

Naming general class

Stressing the qualities which make the object different from other objects in the class

(Function)

e.g.	The lathe	(symbol)
	is a machine	(signifies class 'machine')
	in which work is held so that it can be rotated about an axis while the cutting tool is traversed past the work from one end to the other thereby forming it to the required shape.	(differentiates the lathe from other machines and explains the function)

(*d*) *By operational definition*—Defining by doing. Relating the term to a particular situation.

e.g. The heat in a substance is associated with the motion of its molecules. The hotter the substance the more vigorous are the vibration and motion of its molecules.

(Heat is defined in a specific context and with an indication of experimental evidence.)

Exercises

1. Of which terms are the following sentences intended to be definitions? When you have written down the words, indicate which of the definitions you consider to be suitable. Give your reasons.

(*a*) The process of subjecting objects to extremes of cold for the purpose of preserving them.
(*b*) kWh.
(*c*) The resistance experienced when one body in contact with another slides over that body.
(*d*) A portion of a transparent medium bounded by two spherical surfaces.
(*e*) Power of resisting a change in the arrangements of the molecules.
(*f*) The heaviest of common metals (Sp. gr 11.3); plastic and malleable.
(*g*) The symbol we use to denote that such objects as fresh grass, certain tree leaves, young apples, some gooseberries . . . possess a similar colour, though in varying shades.
(*h*) A force which represents something in a cell which impells electricity through a conductor connected across the terminals of that cell.
(*i*) A device for making a moving object appear at rest.
(*j*) The force which gives a mass of 1 kilogram an acceleration of 1 metre per second per second.

2. State what is wrong with each of the following definitions, then rewrite it in a more reasonable form.

A

(*a*) Length is the distance of one end of a thing from the other.
(*b*) A telescope is an instrument for enlarging objects.
(*c*) Corrosion is the collection of rust.
(*d*) A cell is a supply of electricity.
(*e*) A headstock is the left-hand end of a lathe.
(*f*) Overtime is working late for more money.
(*g*) Temperature is measuring the heat a body has.
(*h*) Grinding is the use of abrasive wheels.
(*i*) A jig helps to guide a machine tool.
(*j*) The point at which a body balances is its centre of gravity.

B

(*a*) Diffraction is the way sound goes round corners.
(*b*) Red is a colour.
(*c*) A proton is a sub-atomic particle.
(*d*) Ceramics is pottery.
(*e*) Prefabrication is when construction work is carried out other than in a place of final assembly.
(*f*) A servo system is a system designed to make your work easier.

(*g*) Refraction is when a ray of light travels obliquely through a different, denser medium.
(*h*) Audibility means hearing above all the other noise.
(*i*) Gybing is turning a boat round.
(*j*) A lubricant is used to reduce friction.

3. The following definitions might be used in specific situations. Comment upon the definitions and say in what contexts you consider they would be useful.

(*a*) Light is a form of radiant energy to which our eyes happen to be sensitive.
(*b*) The velocity of a particle is the rate of change of displacement with respect to time.
(*c*) Sound is a form of vibration. Watch this tuning fork and you will see what I mean.
(*d*) The hydraulic press is an instrument for producing a large force from a small one.
(*e*) Wave motion possesses the characteristic that disturbance travels through a medium without the medium moving bodily with it.
(*f*) Proteins are vital body chemicals and are made from five different substances: carbon, hydrogen, oxygen, nitrogen, and sulphur.
(*g*) A particle is usually understood to mean a very small body whose mass may be considered as concentrated at a point.
(*h*) $P = \frac{W}{g} . a.$
(*i*) When I use the word beauty I intend it to indicate the effect that the general shape of an object has upon the observer's emotions.
(*j*) Static is the colloquial term used to describe the charge of static electricity which accumulates on a vehicle as a result of friction from the tyres on the ground surface.

4. Most of the terms defined in *Exercise 3*, could be defined differently in other contexts. Draft alternative definitions for them and state the context in which they could be used.

5. Provide definitions for each of the following words. Remember that you are defining *words* not things. Use the form of definition which you consider most suitable, and restrict yourself to a single context for each word.

anchor	thrust	telephone
sound	digit	turbine
engineering	torque	oil
lightning	combustion	cone
water	broaching	alloy
rake	micrometer	energy
bonding	plastic	amplitude
momentum	molecule	hydrometer
time	pitch	mineral
chatter	area	architecture
James Watt	calculus	productivity
spelter	creep	derrick
prototype	inertia	gradient
spiral	chuck	fulcrum
shank	filter	spectrum
parallax	calibrate	oscillate
interference	conductor	

SUPPLEMENTARY VOCABULARY LISTS

The following groups of words are arranged under the headings of different specializations. Students should not restrict their vocabulary work to their own section, but should get to know basic words from the others.

Mechanical Engineering

bearing	chamfer	clearance
countersink	capstan	burr
block	coupling	swing
calliper	ribbed	fabricated
cam	web	brazing
lug	apron	backlash
collet	boss	alloy
chatter	lagging	centrifuge
coolant	arbor	lapping
tolerance	leadscrew	broach

profile	reaming	crowning
parting	jig	pinion
skew	land	jennies
plummer	shank	helix
rake	rack	trammels

cotter	spigot	hob
warding	honing	kinetic
thrust	inertia	momentum
synchronous	torsion	alignment
anodize	necking	creep

spindle	chuck	axis
taper	pitch	fillet
platen	witness	turbine
torque	blueing	annular
buffer	crimping	countershaft
reciprocating	follower	eccentric
pivot	collar	journal
sprocket	compressor	isothermal
pneumatic	entropy	scavenge
diffuser	thrust	torque

Electrical Engineering

wave	commutator	ohm
frequency	armature	bega
circuit	brush	bel
hertz	rheostat	decibel
noise	generator	phon
interference	ammeter	lumen
distortion	resistance	coulomb
transformer	conductor	dyne
rectifier	electron	farad
coil	electrode	gauss

electrolyte	potential	attentuation
anode	impedance	absorption
cathode	amplification	compensation
anolyte	impulse	phase
electrolysis	coupling	baffle
electrostatic	trip	batten
electromagnetic	boom	bushing
induction	insulate	fuse
inductance	adaptor	banking
amplitude	burden	core

aerial	screen	solenoid
antenna	fluorescent	vector
dipole	bonding	exciter
co-axial	oscillate	grid

monitor	scanning	admittance
filter	arrester	bridge
balancer	bimetallic	controller
dynamo	field	illumination
feeder	flux	kilowatt
lamination	lead	leak
ion	isolator	loop
lux	resonance	rating
series	static	saturation
mho	dielectric	thermal
thermo-couple	branch	synchronous
conduit	capacitance	shim
interpole	relay	harmonic
divertor	cut-out	sheathing
thermostat	acoustic	polarized
modulation	transducer	thermionic
susceptance	reactance	resistivity
increment	permeability	coercive
hysteresis	photometer	candela
torque	stator	sinusoidal
ripple	permittivity	equipotential
photoconductive	semiconductor	telemetry
slip	invertor	programme
servo-mechanism	slave	diplexer
diathermy	dynamic	oscillograph
equalizer	conjugate	oscillogram

CIVIL ENGINEERING

aggregate	bracing	truss
bond	caisson	bulwark
alignment	aqueduct	buttress
auger	benching	sett
asphalt	crampon	claying
camber	cutback	springing
cant	brooming	adit
grout	coffer	cantilever
flange	derrick	collimation
culvert	triangulation	offset
aberration	dumpling	pile
catenary	gantry	striking
clinometer	spigot	tacheometer

walings	cementation	transit
trunnion	weir	shuttering
rise	springings	pontoon
bowstring	alidade	catchment
cocket	monkey	cribwork
vernier	theodolite	clinometer
sextant	aneroid	altazimuth

dog	dredge	follower
ginging	level	mound
counterfort	chain	pedometer
plummet	orientation	parallax
ranging	resection	staff
tribrach	topography	wye

AERONAUTICAL ENGINEERING

aerobatic	beacon	fatigue
aerodynamics	anhedral	feathering
ailerons	dihedral	astrodome
drift	aerostatics	aerofoil
ceiling	annular	diffuser
duct	cowling	chord
apron	cruciform	camber
approach	gyro	igniter
cantilever	delta	harness
blister	drag	fuselage

fillet	roll	afterburner
homing	slipstream	asymmetric
mode	spar	instability
ramjet	pull-out	supersonic
strut	prototype	ultrasonic
transonic	reheat	subsonic
venturi	geodetic	turbulence
sweep	trim	firewall
thrust	flow	inclinometer
twist	rotor	longeron

drogue	lofting	downwash
hypersonic	fin	pressurized
profile	machmeter	nacelle
overshoot	turbine	vortex
yawing	lift	frame
flutter	airframe	adiabatic

actuator	fail-safe	pod
pitch	stringers	surging
stall	shimmy	rigging
radome	ornithopter	pulsejet

airworthiness	ambient	augmentor
bonding	mock-up	modification
wake	threshold	bunt
rib	tab	cabane
crippling	attitude	radial
safe life	redundant	axi-symmetric
flexibility	diffusion	constraint
stiffness	reaction	coefficient
analogue	determinate	matrix
determinant	computer	digital

Automobile Engineering

carburetter	track	pinion
manifold	induction	helical
compression	cotter	big-end
exhaust	differential	crankcase
suspension	baffle-plate	magneto
tappet	banjo-axle	clutch
hot-spot	boost	camshaft
blow-back	bore	kingpin
blow-by	shoe	shackle
blower	breather	decarbonize

sleeve	synchromesh	louvre
choke	damper	churning
hypoid	wishbone	comparator
hydraulic	understeer	atomizer
hunting	oversteer	dryliner
viscosity	backlash	octane
epicyclic	lock	telescopic
gasket	reseating	port
emulsion	fade	overdrive
priming	additive	sludging

scavenging	degradation	oxidation
capacity	thermostat	torque
strangler	turbulence	inertia
diaphragm	diffuser	venturi
volatile	knocking	drift

stub	pinking	reduction
diametral	worm	cadence
distributor	sheathing	fascia
rimbellisher	baffle	timing
toe-in	inhibitor	splines

Metallurgy

resilient	ductile	malleable	hardie
granular	bloom	sinter	fuller
shortness	ganister	blister	swage
tenacity	ferrite	puddling	flatter
shear	millscale	thermocouple	tup
cementation	refractory	quench	drifts
normalize	arrest	tensile	flash
nodular	ferrous	temper	peen
atomizing	spelter	billet	wrought
blow	occluded	muffle	extrude

Mathematics

integral	abscissa	aliquot
integrate	isosceles	range
rational	ratio	alternate
angle	amplitude	isometric
reduction	resultant	argument
arc	identity	revolution
rhombus	asymptote	acute
latitude	sample	binary
longitude	segment	sector
series	locus	binomial

calculus	logarithm	scatter
sequence	cube	lamina
circle	surd	modulus
circumference	surface	sphere
mean	cotangent	suffix (suffices)
chord	cone	median
mode	similar	scale
cycloid	moment	skew
symmetric	momentum	co-ordinate (coördinate)
combination	numerator	statistics

symbol	cosine	normal
sine	conic	cylinder
obtuse	tangent	ordinate

ogive
triangle
denominator
trigonometry
deviation
trace
parabola

decimal
trajectory
progression
parallel
parallelogram
uniform
ellipse

declination
orthographic
trapezium
differential
diameter
percentage
perpendicular

variation
envelope
polygon
focus
volume
geometry
wave
gradient
projection
pyramid

Pi
extrapolation
vector
prime
function
probability
zero
harmonic
histogram
horizontal

variable
exponential
eccentric
prism
frequency
vertical
graph
hexagon
proportion
index (indices)

integer
permutation
quadratic
quadrilateral
quartic
involute
quotient
interpolation
irrational
scalene